Cybersecurity

Black Hat Pen Testing:
Post-Secondary Graduate Offensive Exploitation & Breaching

Clifton Ray Wise Ed.D.

Author: Clifton Ray Wise Ed.D.

Paperback : 979-8-989923-69-4

Publisher: Clifton Ray Wise

ACKNOWLEDGEMENTS

We like to think we are *self-made*, but in reality, no human entity is self-made, we are products of our environment, from when we are first brought into the light at birth until the day we draw our last breath and move to darkness.

I cannot honestly write a book about security without expressing the necessity of honesty, honor and integrity. To some human entities these descriptors may not seem like much but to some they mean everything, non-compromising straight-up honesty, honor and integrity.

Speaking of such descriptors I would have to start with my dad, Bill Ray Wise for his unflinching and unwavering propensity for truth and honor. If you saw a hundred dollar bill fall out of a man's pocket, would you walk a mile to return it? Well, my dad would have done just that. He always said "When a man is bared all he has is his honesty, honor and integrity."

"That acorn didn't fall too far from the tree". That is what human elements say sometimes when a son or daughter seems to have turned out just like the father. I can truly say my sister, Chandee Rae Wise-Felder didn't fall too far from her father's tree. Her unflinching and unwavering honor and integrity in single handedly exposing negligence, corruption, and malfeasance within a public university and becoming a whistleblower to local, state and federal authorities. She now truly understands the meaning of "standing alone". She was recently asked if she had to do over, going through the trials and tribulations she has, would she blow the whistle again. Her response: "I was voted into that position as staff regent and it was my fiduciary responsibility as a member of my university board and I would respond the same way today."

PREFACE

Welcome to the world of penetration testing and black hat hacking! If you're reading this writing, chances are you're curious about the unseen and most often taken for granted forces that shield and, sometimes, threaten our digital lives. Maybe you're a business leader, a concerned parent, or just a technical enthusiast who wants to understand the basics of cybersecurity without diving too deep into the technical jargon, which generally seems to overwhelm us.

In an age where our personal information, financial transactions, and even our daily routines are increasingly managed through digital platforms, understanding the basics of how these systems can be compromised—and how they can be defended—has never been more important. Although, this writing will delve very little into the defense, and had been written with the "other" side in mind, the offense.

This writing takes you through some of the most basic principles of cybersecurity as they relate to penetration testing, black hat hacking, and the exploitation and breaching of platforms in a way that's engaging, accessible, and free of unnecessary intricacies. I will clarify the jargon, break down the concepts, and provide real-world examples to help you grasp how cybersecurity professionals work to keep our digital spaces safe.

You'll learn what penetration testing is and why it's crucial for identifying vulnerabilities before malicious hackers can breach them. We'll explore common hacking techniques using free programs, the actual programs hackers use in the real world.

Whether you're looking to protect your business, safeguard your personal information, or simply gain a better understanding of the digital security landscape, this writing will equip you with the foundational knowledge you need. By the end, you'll have a clearer view of how penetration testers play a vital role in defending our digital world, through the use of free hacking programs.

Table of Contents

INTRODUCTION

"It is important to understand the true definition of 'Cyber'. Cyber is a word relating to or characteristic of the culture of computers, information technology, and virtual technology. Typically characterized as "the cyber age". Cyber can be a prefix attached to any career designation that previously existed before 'the cyber age". Cyber is most often related to the Internet. The term "cyber" has become ingrained in our language and is commonly used to describe anything related to the digital realm, technology, and online communication."

In today's interconnected world, where a single click can unlock doors to our most sensitive information, understanding the basics of cybersecurity is more important than ever. You might think that hacking and penetration testing are topics reserved for tech experts or IT professionals, but the reality is that these fields are relevant to everyone—from business executives to everyday internet users.

Welcome to a journey into the heart of digital security. In this writing, I aim to demystify the complex and often intimidating world of penetration testing and black hat hacking. We'll peel back the layers of technical jargon and present these concepts in a way that's easy to grasp, even if you don't have a background in technology.

You might wonder: Why should you care about hacking and penetration testing? The answer lies in the growing number of cyber threats we face. As technology evolves, so do the methods used by those with malicious intent. Understanding how these methods work and how they can be countered is crucial for protecting our digital lives.

Penetration testing, often called "pen testing," is a proactive approach to cybersecurity. It involves simulating attacks on a system to find vulnerabilities before they can be exploited by malicious hackers. Black hat hackers use similar techniques as their opponents but with the intent of breaching security through the use of free programs, and exploiting those breaches for illegal purposes.

In this writing, I will start with the basics. You'll learn what penetration testing is and why it's a vital component of a robust security strategy. We'll cover the different types of hacking, from the benign to the criminal, and explain how hackers operate.

By the end of this writing, you'll have a clearer understanding of how to safeguard your digital assets, whether you're a business owner concerned about company data or a curious individual looking to learn more about the digital world.

With knowledge comes the power to protect and defend, and in the world of technology, that's a power worth having.

No Defense Here

In 1999 the National Security Agency (NSA) working with other United States governmental agencies started an accrediting body called the National Center for Academic Excellence (NCAE). In the early stages of that accreditation it was thought that the less the public knew relatively about the offence side of security the better off the nation and the public would be. At that time this accreditation only provided coursework for defensive relationships to computing. In 2012, the thoughts changed from just defensive techniques to adding offensive techniques to the accreditation, as there is today. In my opinion, you are a much more complete cyber-human if you are well versed in the realm of offense and not just a base of defense.

The base of defense is applying patches and updates to computing systems. Not procrastinating and doing them when you feel like it, it is doing them when they become available as soon as you can. Yes, you generally have a test system and a production system in which you install on the test then review it for a length of time, then if all is well, you install on the production system.

What are patches and updates? They are upgrades you might say that fix the holes in systems that have been found and or breached systems. You install the patch or update and the hole is closed or the issue is resolved. This is the heart of defense. I don't see the purpose of a whole course on defense unless I cover all the information related to the operating systems, networks, etc. We will not be spending that time here as it is presumed you already have a decent base of knowledge on those topics.

One more things and I will end here. There are programs and people that sit in a seat all day and watch network traffic pass back and forth through their network looking for entrants who should not be there or looking for malicious human elements trying to get in. We will not be tracing the movement of students within this course.

This course is not the catch-all of penetration testing as there will be no software or human entity tracking your progress in trying to breach these stations. You will not have to worry about being caught in the act and halted. In an actual real-life scenario that would be one of the most important factors of actually being a black hat hacker. The idea that you are anonymous in your trials and that you know just when to pursue an opening and when to back out and regroup.

1 ... G R A N N Y ' S   C O O K I E S

Granny's cookies need to be as safe and secure as anything else. Sometimes we tend to think of things on a large scare to make them applicable. Sometimes we take the smaller things for granted. When we talk of Granny's cookies, maybe this just one simple family's granny or a major company named Granny's cookies. Does it really make a difference? Maybe that is part of the problem as our thinking needs to be more open minded.

With this scenario, does not granny have the same right of her product being protected from harm? When I was a small kid granny's cookies were the best and highly prized by me and my sisters. We would often stop at nothing to get our hands on one of "My Granny's Cookies".

S e c u r e   a s   G r a n n y ' s   H o u s e

Think about the security of a house. A house is ONLY as secure as YOU make it. How much does it mean to you? How much protection are you willing to give? What is the level of risk?

Risk is relative to insurance. The assessment you give for your assets in relation to the insurance you provide to keep those assets from being compromised.

If you own a Volkswagen car you will not pay the insurance premium of a Porsche! Or will YOU? What is the replacement cost of the Volkswagen in comparison of the Porsche?

Back to the house.............

YOU feel safe in your house, right? What precautions have you physically put into place to protect you from the outside?

Let's START with this!

Grandma has the BEST cookie recipe on planet earth! Everyone likes Grandma's cookies! Everyone want's Grandma's cookies!
YOU want those cookies! YOU just gotta have those cookies! YOUR mouth is watering for those cookies!
You know nothing about Grandma's house. YOU have never been in there for 20 years!

D e f e n s i v e   P r o v i s i o n s

What defensive provisions does Grandma have in her house to block the predators from coming in and getting her cookies?

- Does she have a door or doors?
- Does she have locks on those doors?
- Does she have windows?
- Does she have an alarm?
- Does she have a dog?
- Does she live alone?
- YOU take it for granted she has electric and water, you know utilities.
- If she has electric? What defensive measures require electric? Maybe she has an alarm?

- If she has an alarm, does that alarm have a battery back-up in case the electric goes out?
- What kind of alarm does she have, if she has one? Does it have window sensors, glass break sensors, movement detectors, infra-red, etc.?

How do you know? How do you find out?

RESEARCH! INVESTIGATE!

A hacker might wake up one morning and decide to hack something or someone for some reason, but if he/she wants to stay out of jail, they DO THEIR RESEARCH!

ALL it takes in ONE slip-up or mistake at the wrong time to go to jail!

Grandma KNOWS everyone LIKES her cookies! Grandma KNOWS somebody MIGHT want to steal her cookies! Grandma has a recipe for those cookies! Where does Grandma keep the recipe for her cookies? Has anyone ever done a story about Grandma and her cookies? Reporters, newspaper, talk show, etc.

If so, maybe there is information she or they provide during that conversation you can use in your RESEARCH? What was said? What did she or someone else tell? Maybe unknowingly someone divulged information?

RESEARCH! HALF the battle is the Research!

Research

Information is everywhere. In the internet world it is unlimited.

So far we know we like Grandma's cookies. We know Grandma knows we like her cookies. We know Grandma has a cookie recipe. We would take it for granted she wants to keep her recipe safe, right?

How is Grandma keeping her cookies and recipe safe?

People GENERALLY are basically all the same, in theory anyway. People get tired. People get lazy. People sometimes tell things. Sometimes they have a BIG mouth. Many times they think they are safer than they are! Many times they think they and their items are safe enough.

Think about yourself! Are you safe enough? At home, in the car, etc.?

YOU are NO different than the majority of the population!

YOU probably won't realize how vulnerable YOU are until it is too late! This is generally the norm!

The parent let's their kid play in the street! She does NOT think a thing about it! UNTIL the kid gets ran over, and it is too late! Then it is the DRIVER of the vehicles fault! Right? When you think about it: The vehicle and driver had the right to be in the road. That is where vehicles belong. Kids do not belong in the street, right?

Back to our story…

We are talking about a small time scenario. If you think of a LARGE corporation, they are always in the news spouting off about how big they are, and how safe they are, and how they have the most current protections. Always boasting and bragging. If you dig a little you might be surprised what you find in your research.

It does not really matter how small or large the proposed victim is, the procedure and steps are the same.

- Who is the victim?
- What is the victim?
- What is the prize?
- What is the risk?
- Risk compared to Prize?

R i s k

Now you have done your research. Through that research you may have, and probably have come up with even more questions. Questions that need an answer! YOU can NEVER do TOO much research! YOU can never be TOO careful! Prison and jail is NOT a place YOU want to be!

Part of the risk isn't just the obstacles you are trying to overcome.

Has something been put into place to track YOU or watch YOU when you are trying to breach something, a perimeter or network?

If you are walking past Grandma's house do lights come on outside? Are their cameras outside? Can you see them? Maybe you cannot see something that is there? Is it worth the risk?

If you are trying to get Grandma's cookie recipe, where is it? IS it in a computer, a laptop, a server, a database, etc.? Where is it? Who setup the location where the recipe is stored? Was it a layman or a professional? Is there someone hired 24/7 to physically watch monitors and screens to track and watch someone moving around the perimeter of Grandma's house? Or not? If the recipe is on a network, is there someone there watching and tracking network movement and traffic 24/7 in and out of that network? Do you know?

A lot of questions, right?

Are You Interested ?

This is a typical scenario. No matter how large or how small, that does not matter. The scenarios are all the same. It is like that series you watch on television, where the same thing happens every week it just has different participants, a different location, and a different title.

2 ... S E C U R I T Y

It always seems to amaze me when some big time know-it-all comes on the television and talks about how secure things are or seem to be. I always say to myself "why would any schmuck get on national broadcast television and put a target on his back"? That never makes much sense to me. The realism here is that you won't be trying to defend your supposed secure situation against just one hacking entity, you will be possibly defending against hundreds or thousands of want-to-be-hackers looking to make a name for themselves, king of the mountain so-to-speak.

Isn't it enough of a chore to keep something secure to begin with without inviting the world population to try and find your holes or openings? I always tell my students to keep your mouth shut. You may think you are secure and you may think "hey I'm a pen testing student and I know more than you" or "hey, I'm going to college and I know more than you" but, do you really? In this vast world population there is always going to be someone smarter than you, someone bigger than you, etc.

You never know what or who you are dealing with on the other side of that internet connection. Do you really want to risk being another one of those statistics? When you get on television and let the whole world know you exist and you challenge all comers, what do you expect? You may call attention to every Tom, Dick, and Harry that thinks they are a hacker and you surely will draw a crowd of proven hackers.

If you get a suspicious e-mail just delete it instead of making a scene and sending a reply to someone you do not know. Like I said, you never know who or what is on the other end. When I see someone spouting off I always ask them "Are you trying to make us believe what you just said or yourself"? Security is more of a belief, a belief that you have done the most you can do to be as secure as you can be.

D e f e n s e   a n d   O f f e n s e

I see defense as something small and the offense as something way larger. Defense is this little ant and the offense is this myriad of ant lions, if you know what an ant lion is? It is a real opponent of ants. It is the big bully of the ant world along with many other predators of ants.

When I was in junior high school one of the daily classes I had was a construction class. We went several days a week to a building site where we actually built a house from scratch and after several semesters the completed house was sold and the profits went to the school. All of the students in that class rode a school bus to the job site and back each day. After a few months every day there was an issue between one of the bigger students in the class and my friend who was just a simple quiet country boy. The bigger guy was from the city and there always seemed to be something brewing. Well, after quite a few issues between the two, one day we all got off the bus and the big guy approached my buddy and pushed him.

Well, needless to say the fight was on. The big guy kept coming after my friend but could not seem to get hold of him. My buddy was dancing around and swinging both his fists in the air in circles. I don't know where he learned how to do that, but when it was all over the big guy got no punches on my friend and my friend gave the big guy two black eyes. Well, the big guy never bother my friend again, I think he learned some important lessons.

At that same school there was a football jock always pestering this other kid from the country. He was always saying something to him or cutting in line on him, just always trying to push him too far. One day the football guy cornered the country boy in front of what seemed to be the whole school. He told the boy to meet him tonight on the football field after school and they would have it out. The country boy said nothing but everyone showed up on the football field after school. The country boy just stood there as the football played punched him in the face time after time. The boy just smiled after several times and just walked off. The football player was really made by everyone to feel like a jerk. He was provoking someone to fight who had no intention of fighting. A few lessons here to learn too.

A couple more scenarios. I was at a Karate tournament one time and all of my friends were gathered around. One of our friends were fighting. This wasn't a PKA match just a typical match for three rounds and one referee. I have been to many of these tournaments and it seems you always have the show off guy and the quiet guy. It seems most matches are that way. I don't know why, they just seem to work out that way. One guy tries to scare his opponent or psyche-him-out then the other guy just looks down and tries to pay little to no attention to the hype. This was one of those situations.

My friends was the quiet guy who was just happy to be in the ring and compete. The first few rounds when the referee would start the round the bigger guy would charge at my friend swinging wildly and just about push him out of the ring. If you get pushed out it is a point for the other side. The ref would then shout "point" then point to whoever earned the scored point. After two times my friend was losing. This big bully was just flailing his arms wildly and basically pushing my friend from the ring. I told my friend the only way he was going to win this match was is if he used the guy's posture and weight against him. At this point it was two points for the opposition and zero points for my friend, out of a possible three points to win.

The big guy came charging, my friend stepped aside and let the big guy basically pull himself out of the ring. One point for my friend. The big guy had a look of surprise on his face. I told my friend let's really surprise him this time. The guy came charging, my friend turned 180 degrees, bent down with his hands and touched the ground, then kicked his foot up against the chin of his opponent. Another point for my friend. Was he really surprised this time, yes he was. The last part of this bout was for my friend to rush his opponent as soon as the call was made to start. My friend rushed up, and pulled him in. Point three for my friend. Some lessons learned here.

At another Karate tournament. All styles were invited to compete. Fighting styles that is. For several years there had been a new studio (dojo) in town and there they were taught Kung Fu. One of my friends was one of the students there. My friend was already a Black Belt in Shaolin and wanted to learn another style. The instructor was a master and champion from around the world and not well known in our area. This instructor brought several of his students and came to the tournament and asked to be admitted and allowed to compete with all the other styles. He was told no, and that the reason was this was a fighting tournament and Kung Fu was not a fighting style. This instructor did not take it nicely! He stepped into the ring and challenged all participants at that tournament.

He said "I will show you a fighting style, if anyone in this tournament has the %^$#@ to face me"! Of course, when you have an enormous amount of testosterone in a building you will have at least one taker who thinks he can conquer the world. One candidate stood up and entered the ring. The Kung Fu instructor hit him three times before he could even blink and he was face down in the ring. The Kung Fu instructor had some choice words to say and this issue has never been resolved even up to this day. Lessons, you bet.

W h y   t h e   M e n t i o n ?

I see defense as the small guy who is trying to protect himself or something and the offense is the larger bully type guy who tries to breach. Life is full of these instances. We will always have David and Goliath scenarios, good trying to overcome evil and vice-versa. These are daily occurrences since we live in a non-perfect world. If the world was perfect we would not have to worry these things would not happen.

I always hated the saying "It is what it is". Whomever came up with that looked at life as limited. Sure life has limits but for the most part we as human entities with logical brainpower can pretty much change what we want to change. Human entities put limitations into place. Most of the time limitations can be changed or are flexible. We just seem to for the most part make things or see things as more complicated than they are. Once we study and review a bit we say "ah, that isn't so hard". "It is what it is" is the predetermined theory or concept of a closed mind that every situation or instance that particular human entity did their best or all they could do at that time. That to me is ludicrous in itself. We as human beings, some of us anyway, try to do our best, most or all of the time, and sometimes our best just isn't good enough or there is something we left out causing further issues or complaint.

We have to think of everything we do as a job. No matter what color hat hacker you are or think you are or study to be you are doing a job or at least trying to do a job, or at least a task. The black hat, gray hat, and white hat are all doing the same thing, they just go about things a little differently, the white goes by rules and regulations, the black has no rules, and the gray knows and thinks about the rules and regulations and comes up with justifications within his/her own limits and minimizes the effect or affect. It is just a job.

The morals of this chapter are simple. There are defenders and offenders. Know your opponent. Don't take anything for granted. The stories above could have just as easily turned out the opposite, they just turned out that way. Was it preparation or just time and circumstance? Good things happen to bad people and bad things happen to good people. Do the best you can and sometimes, sometimes, it may not turn out to be good enough or have the outcome you were looking for when you began.

There are TWO types of people (human entities) currently on this planet.

When something BAD happens to each of us, what do we/you say?

1. I did what I thought was right, Why did this happen to me?
2. I knew it was coming, it was just a matter of time!

Who do you want to be?

W h a t - I f

When we talk about security and all the relationships we understand honor and integrity to be key elements. No matter what we do or how we carry those out, we have to plan ahead and usually we, as the human element have to plan ahead by looking at the past, past issues that may have compromised our supposedly secure situation.

In my opinion you can never plan enough, but we have to limit somethings sometimes, there just isn't enough hours in a day or enough money to project top keep things totally as secure as we think them to be. Sometimes there are what I call "what-if" scenarios. We can never think up enough what-if scenarios and that si not the point I am trying to make. The point I am trying to make is, no matter how hard you try you cannot plan enough for everything, it is just not feasible. In security we have Business Continuity Plans (BCPs) and Disaster Recovery Plans (DRPs) and those are crucial, but however you peel them apart you cannot plan for every possible instance that may come your way.

What-if a clown came to your door with an axe and started chopping at the door to get in? Have you planned for that? What-if an elephant came into your yard with a ninja dressed in black had a machete and tried to pry your window open? Have you planned for that? I know, they do seem a bit extreme and not possible, but the point is, in this day and time anything is possible. Could they in fact happen? You bet they could. Could a ten year old or eighty year old human try and breach your network? You bet they could. Just because these things are not on your radar as you have never came across them in your existence on this planet does not mean they do not exist or could not happen.

In Closing

- Plan as best you can
- Don't put a target on your back
- Don't invite response
- Don't think it cannot happen to you

I was driving down the road one day and my bank called. They asked when the last time I was in Canada and if I was back home at that time, and how I enjoyed my trip. I said I had never been to Canada. They said "apparently someone had my bank card and started on the west coast of Canada and travelled all the way back to the east coast and was using my card extensively". I opened my wallet and said my bank card was with me personally and I had it all along in my possession. I don't remember how many thousands of dollars were spent on my card. I do remember the ordeal of affidavits I had to file and complete to get those changes removed from my account and reimbursed to me. I remember the hassle it was and the months it took to get my identity back. It wasn't something I care to do again and you will not want to either. My grandmother had this happen to her also. What is the statistic now, one in four human beings will have their identity compromised.

3 ... THE PEN TESTER AND BLACK HAT HACKER

In this chapter we will discuss the penetration tester, also known as the Ethical Hacker, and the black hat hacker.

Pen Tester

Pen testers are authorized to test and evaluate the security of systems. They are contracted to work within the boundaries of an organization's consent and legal agreement. The goal of the pen tester is to identify and report vulnerabilities, to recommend fixes, and to help improve the overall security position of an organization. The scope of the pen tester is to operate within a predefined scope of work (SOW) agreed upon by an organization, ensuring that their activities do not exceed the boundaries of their authorization.

Pen testers have proficiency in a multitude of various cybersecurity tools and techniques, programming languages, and the understanding of security frameworks in which they work. Pen testers document their findings by writing detailed reports and communicating those recommendations to both potential the technical and non-technical stakeholders that hired them.

Pen testers must adhere to ethical guidelines and legal standards, ensuring that their work is conducted with integrity and respect for privacy. Pen testers conduct what are called "vulnerability assessments". They assess the potential vulnerabilities within a system or network by seeking out weaknesses. They may also perform social engineering tests and or the testing of web relative applications.

B l a c k   H a t   H a c k e r

This is NOT to be confused with the Ethical Hacker! The Ethical Hacker is another name for the Penetration Tester mentioned above.

The black hat hacker is the entity that uses any means necessary to search out a hole and breach it. They are not tied to any rules or regulations, stipulations or specifications. It is their position to breach a system or network by any means necessary and to exploit that breach to whatever motive or purpose they desire.

The goals of black hat hackers can vary from a multitude of illegal activities. Black hat hackers operate outside the laws, rules, and regulations of society without authorization, and typically engage in activities that are harmful and or illegal. Black hat hackers possess strong technical skills, however, their techniques and tools are most certainly used for illicit and illegal purposes. Depending on their objectives, black hat hackers use various methods, from exploiting vulnerabilities to social engineering and or the crafting of sophisticated malware themselves.

Black hat hackers operate outside legal and ethical boundaries, engaging in activities that can cause harm or breach privacy. They conduct scans and engage in unauthorized access to systems or data, they may deploy ransomware or other forms of malware, and they may engage in phishing scams or other deceptive practices.

Comparison of both

Penetration testers work with explicit authorization from the organizations they are testing. Black hat hackers, especially malicious ones, do not have such authorization and often operate illegally. Penetration testers aim to improve security and help organizations by identifying and fixing vulnerabilities. Black hat hackers' intentions can vary widely from malicious activities to constructive ones, but they often lack the legal and ethical considerations of professional penetration testers. Penetration testers operate within a defined scope agreed upon with their clients. Black hat hackers may act without any scope or boundaries, often causing harm or breaking laws. Penetration testers adhere to legal and ethical standards, while black hat hackers do not.

In Closing

While penetration testers and black hat hackers may use similar technical skills and techniques, their purposes, authorization, and adherence to legal and ethical standards distinguish them significantly. Penetration testers are professionals who work within the boundaries of the law and ethics to enhance security, whereas black hat hackers engage in illegal and harmful activities.

4 ... TRAITS OF A PEN TESTER

A penetration tester must have a diverse set of skills and characteristics to effectively identify and address security vulnerabilities. A penetration tester is basically a black hat hacker that follows ethical rules, regulations and guidelines. They follow a specific path and do not deviate from that ethical path as in the case if the black hat hacker.

Programming

Pen testers must have proficiency in various programming languages such as Python, JavaScript, and or C/C++. They also need to understand the behind-the-scenes workings of computers and computer networks. They need to understand the workings of operating systems, platforms, and protocols as well as the myriad of tools and techniques used in penetration testing.

Problem - Solving

Pen testers must have strong problem-solving abilities and analytical thinking skills to understand complex systems, identify vulnerabilities, and determine how they can be exploited. Pen testers must meticulously analyze systems and networks to spot subtle vulnerabilities, misconfigurations, or open holes that might be overlooked in a typical computing environment. Pen testers need to think outside the box to find and exploit weaknesses. Creativity in developing new attack strategies and methods is a valuable trait.

P e r s i s t e n c e

Pen testing is most often challenging and time-consuming and persistence is generally necessary to thoroughly test systems, overcome obstacles, and find hidden vulnerabilities. Adhering to ethical guidelines and legal boundaries is crucial in a pen tester as they must work within the scope of their legal agreements and respect privacy and confidentiality. Pen testers must clearly and effectively communicate findings, both in writing and verbally, including the creation of detailed reports, the explanation of vulnerabilities to non-technical stakeholders, and the providing of actionable recommendations for remediation if necessary.

D e v e l o p m e n t

Pen testers are committed to ongoing education and professional development by staying updated with the latest threats, vulnerabilities, tools, and techniques. Pen testers must be open to and adapt to new technologies, changing environments, and evolving security landscapes, including adjusting testing approaches based on different systems and client needs. Pen testers can work independently in some instances, although they are often required to collaborate with other team members, clients, or IT staff.

Envision

Pen testers must understand how to assess and prioritize risks and the potential impact of vulnerabilities by helping clients focus on the most critical aspects of their environments. Time management skills are necessary to balance multiple tasks, meet deadlines, and manage the overall scope of a pen testing assignment. A genuine curiosity about how systems work and a passion for discovering vulnerabilities drive many successful pen testers. This trait often fuels continuous improvement and innovation in testing methods.

Solving

The ability to troubleshoot issues, think critically, and develop solutions to overcome technical challenges is important in a pen tester. Familiarity with security frameworks, methodologies, and standards (such as OWASP, NIST, and PCI-DSS) can be beneficial in structuring and conducting thorough tests. By embodying these traits, a pen tester can effectively assess the security of systems and help organizations improve their defenses against potential threats.

5 ... TRAITS OF A BLACK HAT HACKER

Black hat hackers come with different motivations and methods. While there isn't a single profile that fits all black hat hackers, several traits and characteristics are commonly associated with them.

Black hat hackers are often driven by a deep curiosity about how systems work. They enjoy exploring and understanding the inner workings of software, hardware, systems and networks. They generally possess a strong knowledge of programming languages, operating systems, networking, and cybersecurity principles. Black hat hackers frequently have advanced technical skills and stay updated with emerging technologies and vulnerabilities.

Black hat hackers are typically skilled problem solvers who can think critically and creatively to exploit and breach weaknesses in systems and networks. They often enjoy solving challenges and mysteries. Numerous hacking scenarios require significant time and effort, therefore, black hat hackers often display persistence and tenacity in pursuing goals, whether that goal is gaining unauthorized access to a system, network, or device or developing new feats.

Success in hacking often depends on a meticulous approach to identifying vulnerabilities and understanding complex systems. Black hat hackers must pay close attention to small details that could indicate a security flaw or potential holes. Black hat hackers are highly aware of the need for secrecy and anonymity and value those. They generally use techniques and tools to hide their identity and location, such as VPNs, proxies, and encryption.

The technology landscape is constantly evolving, therefore black hat hackers must be adaptable to quickly learn and respond to new technologies, security measures, and defensive techniques. Some black hat hackers participate in communities or forums where they share knowledge, tools, and techniques with other white, gray, or black hat hackers. They often collaborate with others to develop new exploits or achieve common goals. They may often choose a victim and make a game out of breaching that victim.

Hackers are generally motivated by various factors, including financial gain, political and or ideological beliefs, personal challenge, revenge, or simply the thrill of breaking into a system. Understanding these motivations can provide insight into their actions and strategies. Most often engaging in hacking related activities often involves significant legal and personal risks, whereas those black hat hackers may need to possess a high tolerance for risk and be willing to face potential legal consequences if caught.

Black hat hackers often have a strong desire for continuous learning and self-improvement as they must keep up with new hacking techniques, vulnerabilities, and cybersecurity trends to potentially sway their being captured. Black hat hackers must be resourceful and innovative to achieve their objectives.

Ethical hackers or pen testers certainly have a similar skill set to black hat hackers although they apply their knowledge to help organizations improve their security, and the black hat hackers use that same knowledge to exploit and breach.

Understanding these traits may help in developing effective security strategies and recognizing potential threats.

What Exactly is a Hacker Anyway?

Remember talking about White, Gray, and Black hat hackers (Hack3rz)?

Generic Image: Three Hats of Hackers

Black hat are the bad guys!

Gray hat are the guys who think they may be doing something for the good, but they are still breaking the law!

White hat are the good guys!

My Proposed Colors

Where I propose the colored names of the hats came from? In the 1920's, 30's and 40's in the USA we had what were called 'Serials' and the majority of these were 'Serial Westerns'. This is where some of the movie stars today got their starts in showbiz: John Wayne, Tom Mix, Hop-along Cassidy, etc. In the first years the good guys wore white hats, the bad guys wore black hats and the guys who could go either way wore gray. In the following years the hats changed colors and those colors were not as prevalent.

It would not be complete if I did not mention Crackers (Crack3rz).

Software Crackers

In years past Crackers were more prevalent. You know, back when we had dial-up internet connectivity. At that time, you did not have an internet connection that was unlimited. At that time your connection was limited. You used a dial-up modem to connect to the internet, you conducted your research, checked your e-mail, did a little looking around, and got off the internet. Now, you probably leave at least one computer or laptop at home connected to the internet all day, while you are away? Do I need to express to you how that is NOT a good idea?

Anyway, have you ever bought a piece of software and after you installed it on your PC you had to include the serial number to enable and validate the software? You validate the software to say that is NOT a pirated version, it is your real purchased copy, right? In the old days programs were stand-alone and now they are not. Now when you install software on a PC it automatically updates that serial number that is listed in a database and validates your software program. It does this because you are probably connected to the internet full-time. In the days gone by when Crackers were more prevalent, the Crackers would 'crack' the serial number of the software program and spread that serial number around to everyone else. They could do this as companies at that time did not have a way of connecting to that software you purchased unless you were on a full-time or part-time internet connection.

You could use the cracked serial number as long as you never connected that PC to the Internet. Those days nave since passed. Now some software programs run on servers or what we call the cloud to give consumers access and to control that your software is valid.

This newer 24/7 connectivity of PCs, laptops, workstations, servers, network devices, cell phones, and wireless has opened up a whole new era of "Un-Safe"!

6 ... PEN TESTER MISTAKES

Don't get confused here! This is not a comprehensive list of exploits related to Hacking! This is a listing of breaches relative to the exploits of the Penetration Tester. Let me put this another way! If, the Pen Tester does NOT do his/her job correctly THESE are the open avenues of their mis-attempt at resolve.

Beyond

If a penetration test is not properly scoped or managed, testers might accidentally gain access to sensitive data or systems beyond the agreed-upon scope, potentially leading to security breaches or data exposure. Poor communication between the penetration testers and the organization can lead to misunderstandings about the test's scope, objectives, or limitations. This might result in unintended disruptions or damage.

Sensitive

During a penetration test, sensitive data might be exposed or collected. If this data is not handled securely, it could lead to data breaches or leaks. Penetration tests can sometimes cause unintended disruption to systems or services, which might impact business operations. If not managed properly, this disruption could lead to operational downtime or loss of service.

Strategy

Relying solely on penetration testing without a comprehensive security strategy might give a false sense of security. It's important to integrate pen testing with other security measures like regular patching, monitoring, and user training. Test findings, if not handled carefully, can be leaked or disclosed prematurely. This can potentially inform attackers about the vulnerabilities before the organization has had a chance to address them.

Ethics

Penetration testing must be conducted within legal and regulatory frameworks. Any deviation from these frameworks can lead to legal consequences or compliance violations. If penetration testers do not provide thorough and clear reporting, the organization may not fully understand the vulnerabilities or the necessary remediation steps, which could leave security gaps unaddressed.

Credentialing

Testers often need to use credentials to access systems. If these credentials are not handled securely, they might be exposed or misused. If the penetration test does not include social engineering assessments, it might miss vulnerabilities related to human factors, which can be a significant security risk.

To mitigate these risks, it's crucial to carefully plan and scope the penetration test, ensure clear communication with all stakeholders, handle data responsibly, and integrate penetration testing into a broader security strategy.

7 ... WHAT IS A 'THINKING HAT'

In this chapter I am going to talk about thinking. Many years ago the thinking was right-brain and left-brain. It was though that human elements had one or the other as dominant. You were either a creative or critical thinker, not 100% but more of one than the other. It sounds logical, right? Some of us are better at being creative and some of us are hard and deep thinkers, right? Then some of us think we are just the "cat's meow" at everything. My question would be: What does everyone else think about that?

I like to think of creative and critical thinking as working together. Kind of like a "Yin and Yang". Both things make you complete. I like to think of them as two brothers working together to solve a common goal. Then we have this thinking method called strategic thinking. Strategic thinking is like the cousin of two brothers. You know the third wheel when you get together, the guy who can't make a decision without looking at both sides of all issues, you know the politically correct thinker that "beats the dead horse" so-to-speak. You don't mind him being around or coming to the party, you just want him to get to the point!

Thinking is a series of processes we do not realize, as happening every second of every minute of every day. Pulses bouncing from left to right and right to left in our every waking moment as well as when we are at rest. Some thoughts we control, or try to control, and some thoughts we may not even recognize are being processed through our subconscious, while sleeping.

The human mind is always trying to *work things out*, to *decipher*, to *interpret*, and to *untangle* pieces of data the physical senses bring to its attention: sight, smell, hearing, taste, and touch. The senses open up a door of awareness to our minds. Once the door is opened our mind starts to analyse and process what just happened, and we usually don't even realize it is happening, it happens sometimes in the background. This is the *blank slate* each human is born with that starts to compile information form the first day it is brought into the light and continues until the final day when that light has extinguished.

During the lifecycle of this *blank slate*, foreseen and unforeseen issues and problems abound over the years. Issues that can be small or large, and simple or complex. Issues that are either under our control and or not of our making but that require our input, for a solution. When a problem or issue is sensed the brain starts to go through analysing and processing the received data.

A solution for this sensed problem or issue is limited by the current knowledge base of the recipient, and in fact each human entity is a product of its environment and we all have limitations of previous experiences and familiarities to help with our resolve.

There are two sides to the human brain, the left and the right. Both sides work together sending pulses back and forth. Some people are left-brain dominant, some are right-brain dominant and some use both-sides somewhat equally, which is a lesser percentage than the left and right dominant. People who are labelled as left-brain thinkers are said to be logically and analytically dominant. People who are labelled as right-brain thinkers are said to be innovatively and creatively dominant.

I think you get the point, enough about that, let's move on. The proceeding pages are relative to the left-brain dominant, logical *critical* thinker and the right-brain dominant, innovative *creative* thinker and the thinker that combines the two, the strategic thinker, maybe even doing this unknowingly. Problems and issues, are sensed and analysed by the brain where it decides what tools are needed to help resolve the situation. Remember now, each human has a limited knowledge base, limited by the experiences of what they have learned and acquired from each of their respective environments. Sometimes we need to use our logical side for resolve and sometimes we need use our creative side. Some of us are better at some things and others other things. Sometimes we need to think logically for a solution and sometimes we need to look at being creative.

'Ten Gallon' the Critical Thinker

Why the Ten Gallon Hat in relation to Critical Thinking? The Ten Gallon Hat is deep, and critical thinking is often thought of as deep or profound thinking. We'll call the Ten Gallon Hat the Thinking Hat, the Critical Thinking Hat. When I think of critical thinking, I think of logic or thinking logically. That may be a stretch to others in the field, but that is how I see it. I see it as looking at an issue or problem and realistically looking for answers to solve it, as your brain scans through the folders of knowledge you have attained throughout your short existence to find a solution. What process does the human mind need to go through to solve that issue or problem? Is it a small process with only a few steps, or is it a larger problem that requires many steps and/or sub-steps within the steps? This is what I derive as the Ten Gallon Critical Thinking Hat

As with the critical thinker, there are many ways to articulate the concept of their thinking, critically that is. Yet every substantive conception of critical thinking must contain certain core elements. Consider the following brief conceptualizations.

"Critical thinking is the intellectually disciplined process of actively and skillfully conceptualizing, applying, analyzing, synthesizing, and/or evaluating information gathered from, or generated by, observation, experience, reflection, reasoning, or communication, as a guide to belief and action. In its exemplary form, it is based on universal intellectual values that transcend subject matter divisions: clarity, accuracy, precision, consistency, relevance, sound evidence, good reasons, depth, breadth, and fairness..."

"Critical thinking is self-guided, self-disciplined thinking which attempts to reason at the highest level of quality in a fair-minded way. People who think critically consistently attempt to live rationally, reasonably and empathically. They are keenly aware of the inherently flawed nature of human thinking when left unchecked.

They strive to diminish the power of their egocentric and socio-centric tendencies. They use the intellectual tools that critical thinking offers – concepts and principles that enable them to analyze, assess, and improve thinking. They work diligently to develop the intellectual virtues of intellectual integrity, intellectual humility, intellectual civility, intellectual empathy, intellectual sense of justice and confidence in reason. They realize that no matter how skilled they are as thinkers, they can always improve their reasoning abilities and they will at times fall prey to mistakes in reasoning, human irrationality, prejudices, biases, distortions, uncritically accepted social rules and taboos, self-interest, and a vested interest."

"Critical thinkers strive to improve the world in whatever ways they can and contribute to a more rational, civilized society. At the same time, they recognize the complexities often inherent in doing so. They strive never to think simplistically about complicated issues and always to consider the rights and needs of relevant others. They recognize the complexities in developing as thinkers, and commit themselves to life-long practice toward self-improvement. They embody the Socratic principle: *The unexamined life is not worth living*, because they realize that many unexamined lives together result in an uncritical, unjust, dangerous world."

Within the industrial working environment, this can be as simple as working on an assembly line and putting nut 'A' onto bolt 'B'. How many steps do you need to draw or spell out to do that simple job or task? Can just anyone or everyone do that simple task? Does everyone need the same number of steps to complete that task? Does the human mind work in all matters the same for everyone? Think about it. Do you think the same way that everyone else around you does? That is doubtful, as if you know anything about the human mind, you know that everyone is different. If all the writings throughout time are right, then each human entity or in relation to the title of this reading, the college freshman, is a product of their environment and surroundings. If correct, these are not the same for everyone. Every potential student is different.

Then there comes the additional factor of age: the age of the college freshman. If we look at statistics in relation to the college freshman, the majority of that population have just graduated high school or have graduated a few years earlier, generally a population between the ages of eighteen to twenty-two. The older we are and the more time we have spent breathing and accessing the realizations throughout our lives, the more blocks of knowledge we have acquired and retained. Our minds retain those ideas, concepts, and solutions, whether we realize it or not. From birth, the human mind is like a sponge soaking up every bit of knowledge your physical senses pick up. Some human senses are stronger and keener than others. After all, we are all different in our makeup, Right? A newborn child has often been spoken of as a *clean slate*: a *clean slate* that progressively soaks up everything it hears, sees, and feels, leading through the years to what we are as we grow older.

Let's look at a more intensive/extensive example of necessary steps to achieve a solution. What if you are thinking of an education and the educational pathway is something within the realm or area of computer science and/or information technology? I choose these realms as these are what I have been involved in throughout my existence on this planet. What issues or problems might a computer science or information technology student run into during their pathway to educational success in the classroom?

Let's first look at computer science. First off, computer science is one of the most difficult post-secondary educational programs that exist today; it is likened to engineering and the sciences. This is not to say they are the only difficult educational pathways; these are the ones I am familiar with and have knowledge of. This is not to deter you, but rather to prepare you. The career positions tied to these programs of study can be highly lucrative and pay very well. That is generally why a lot of potential students seek out these programs. If we are seeking out these programs, we need to understand what is required to be successful in them. After all, we all want to be successful, right? The purpose of getting an education is to graduate and get a good career, right? And of course, make good money, right? Why not reap the benefits that come with it, right?

Within computer science, there are a lot of programming courses, and some are statistics-related. These courses generally take more preparation and mindful thinking to process: to process and successfully work through the problems you may be given in your coursework to find correct and efficient solutions. These problems will not be as simple to solve as the nut 'A' on bolt 'B' example mentioned earlier. These will be more complex with multiple steps and sub-steps. Pages of programming languages and code to work through, to troubleshoot, and to develop.

How about information technology? When you think of information technology, you think of computers, hardware and software for computers, the networking of computers, and the security of all these things. There is a lot of memorization of the concepts and ideas within this coursework, as well as steps and sub-steps of troubleshooting devices and the planning and design of computing networks. Additionally, you will also be involved in programming related to computers and networked devices as well as website languages. These can be somewhat complicated, but they are all guided by the student and their desire to achieve and succeed. The object is not to dissuade you from seeking these programs but to inform and prepare you. You will surely come to use your Ten Gallon Hat. It will become your second nature.

'Sombrero' the Creative Thinker

Why the Sombrero in relation to Creative Thinking? The Sombrero has a wide brim: a wide brim that covers the possible vastness of thinking *outside the box*. Inside the box, *thinking* has the limit of the internal size of the box: the diameter, the circumference, and dimensions of the box.

We'll call the Sombrero the Creative Thinking Hat. Some of us are more 'creative' than others. Some of us are better artists than others. Some of us design, draw, paint, and illustrate, and some of us just don't seem to have that talent. Can you acquire that talent, or is it something you are either 'born with' or not? Well, in previous wording, we mentioned that human entities and authors throughout time have said babies are thought to be born with a *clean slate*, right? That is what has been said. If that is true, then that 'baby' must be trainable or teachable in some form or fashion, right? If the human entities of the world are correct, then it must be.

You just have to come into those situations where your mind needs to view what it sees in a different manner.

Looking at an issue or problem through a different *lens* than you normally see. Have you ever heard someone say "they are looking at that with rose-colored glasses?" What does that mean? That means that you are looking at something with bias. That means you are looking at something without an open mind; you are only seeing what you want to see. You are not looking past your own knowledge base. You are only trying to solve that issue with the knowledge you have attained so far through your life and not looking past that, as you may not know everything and there may be more knowledge you can attain past the point of today, at any given moment in time. Does that make sense?

The creative thinker continually digs deep into oneself and generates more, newer, better, faster, cheaper, different ideas that can be used to improve the important parts of their life, i.e. the successful manager. The creative thinker is comprised of seven distinct qualities:

1. I am curious to a fault
2. I practice zero-based thinking
3. I am willing to change
4. I am goal focused
5. I am willing to admit when I am wrong
6. I do not know everything
7. My ego will not be bruised if and when I am proven wrong

The first quality of creative thinkers is that they are curious to a fault. They are always asking questions. They never stop asking questions, and would like to find others similar to themselves. They ask questions like "Why or Why not, why can't we do that, if it hasn't been done, can we do it now?" There onslaught of questioning would be similar to that of a child, but they know why they ask, they have reasons, other than the bantering of a child.

The second quality of creative thinkers is that they practice "Zero-Based Thinking" throughout their daily routines. They continually ask themselves, "If I were not now doing what I am doing, knowing what I now know, would I start?" If their answer is no, they stop that train of curiosity , then move on to another thought project and start their questioning all over again. They say that hindsight is always 20-20. Creative thinkers move forward and do not dwell on the "what if" scenario. So many humans persist on questioning and spending vast amounts of energy on projects that when they look behind they would have never started to begin with, and generally wish they had not, then they wonder why they make so little progress and it seems to take forever.

The third quality of creative thinkers is that they are always willing to change, they are open to new suggestions and have an open mindset. They prefer to be in charge of their lives rather than being caught up in a flash flood of change that may often be inevitable and or unavoidable. The words of the truly flexible person, the person who is willing to change are simply, "I changed my mind." According to researchers, fully 70% of the decisions you make turn out to be wrong in the long run. This means that you must be willing to change your mind and try something else the majority of the time. Mental flexibility is the most important quality that you will need for success in this the 21st century.

The fourth quality of creative thinkers is their willingness to admit when they are wrong. They do not hold fast to their opinions when they are proven wrong. Researchers say that 80% of people burn up most of their mental and emotional energy defending against admitting that they made a wrong decision. True creative people are open minded, fluid, flexible and willing to both change their mind and admit that they are wrong when their earlier decisions turn out to be incorrect.

The fifth quality is that creative people can say, "I don't know." They recognize that it is impossible for anyone to know anything about everything, and it is very likely that almost everyone is wrong to some extent, no matter what they are doing. So when someone asks them a particular question that they don't know the answer so they admit it early and often. They simply say, "I don't know." And if necessary they go about finding the answer. Here's an important point. No matter what problems you have, there is someone somewhere who has had the same problem and who has already solved the problem and is using the solution today. One of the smartest and most creative things you can do is to find someone else, somewhere, who is already implementing the solution successfully and then copy him or her. The smartest person is not necessarily the person who comes up with the idea, but the one who accentuates or successfully builds upon it.

The sixth quality of creative people is that they are intensely goal focused. They know exactly what they want. They have it written down very clearly. They visualize it on a regular basis. They imagine what their goal would look like if it were a reality today. And the more they visualize and imagine their goal as a reality, the more creative they become and the faster they move toward achieving it.

The seventh quality of highly creative thinkers is that they have less ego involvement in being right. You will not bruise or crush their ego if you prove them wrong. They are always looking for and willing to accept the correct answer even if it down not come from them. They are more concerned with what is right rather than who is right. They are willing to accept ideas from any source to achieve a goal, overcome an obstacle, or solve a problem.

The most important part of creative thinking is the ability to generate ideas. And the greater the quantity of ideas that you generate, the greater the quality your ideas will be. Similar to that of a brain trust: a gathering of multiple brains coming up with multiple ideas to successfully come up with a solution. The more ideas you have, the more likely you are to have the right idea at the right time. But generating ideas is only 1% of the equation. As Thomas Edison once wrote, "Genius is one percent inspiration and 99 percent perspiration."

Your ability to come up with an idea, to test it and validate it, and then to implement it through creative thinking and execute it in your life to achieve results is the true mark of a successful IT manager. Every single time you originate a new idea, write it down, make a plan for its implementation through creative thinking and then take action, you are behaving like a genius. And the more you manage your creativity in this way, the smarter you will become. And the smarter you will become, the more you will achieve in every area of your life.

It is a rather drawn-out and lengthy concept. Basically, it means not to let the limits of what you have learned *soaked up* so far in life deter you from being able to *soak up* new things. After all, none of us know everything; we are only human entities and we are always learning, even when we grow older, we never stop learning. You may think you have stopped learning, but remember in previous paragraphs we talked about the human mind as *soaking up* as it goes along. Well, until we pass through this life, the thought is that we always gather information whether we realize it or not.

Enough explanation and background, let's get to the title at hand: Creative Thinking. Looking at what we need to do and the steps we may need to make to solve an issue or problem creatively. Could we call this thinking 'outside the box'? What does that mean? That means to find a way to solve the issue or problem at hand by using something other than the normal process and by using some mental tool or reasoning each of us may have attained throughout our lives, each one of us separately. Remember, when we talked earlier about the knowledge each of us attains throughout our daily lives and the environments we each are involved in? Remember that? That is what we are talking about. Since we all are a product of our environments since birth, maybe you have come into contact with something to help you personally solve the issue, problem, or task given. Then again, maybe not. Maybe you have not been privy to such an environment and have been somewhat sheltered or are quite young and have not experienced such situations in your limited past?

If I look at this through faculty eyes, we are often given the task of creatively finding a way to reach students. Past history shows educators that the more *senses* we can reach within a student, the more apt we are to give them an understanding of the material we are trying to convey. This brings to mind an example from my past when I was in an Associate Degree program in Computer Electronics. In one of the courses, we each bought a kit. A kit is a lot of pieces of electronic and electrical parts. These combined parts were an AC to DC voltage regulator power supply. We each bought a kit, and the object was to follow the directions and solder up and wire the kit. This kit would be used as our personal equipment in our future higher-level electronic courses. The day came when we were to plug in our power supplies for the first time to see our new inventions light up and get a passing grade.

Before I get to the *punch line*, I will give a little explanation. I transferred to this college from a Flight school in the south. At that Flight school, I was in a highly disciplined environment. At that college, I was one of the top students in my class and was often prodded by my instructors. All the instructors were ex-military, and all students towed the line. Since my father was a drill instructor in the Air Force, and that is how I grew up, I knew my place and my responsibilities. Well, if I wasn't in class at 7:00 am in the morning, the instructor would come beat on my dorm door until I answered. To make a long story short, this instructor made sure I was the best I could be.

The best I could be was to make mostly B's and some A grades for courses. My coursework was beaten into my head at that facility, and when I transferred to another college up north, I made all 'A' letter grades on everything! And I mean everything! Thanks to my previous instructors and their pushing me to be my best! At my new college, I was quickly made the course mentor and assistant for all my electronic/electric courses. I was the kid that helped the struggling students.

Back to the story and the punchline. I plugged my power supply in and passed the class without a hitch. My best friend plugged hers in and BANG! Smoke filled the room! It was like an old Batman serial rerun! Bang! Boom! Pow! But with the addition of a lot of smoke! And the smell was horrible! Have you ever smelled burning electronics? Well, needless to say, it wasn't pleasant, and it does have a smell all its own. Well, nobody got hurt as each power supply was in a metal case with knobs and leads. A good thing, right?

My friend was heartbroken. Something within her power supply exploded. She opened the case, cleared the smoke, and looked to see what had gone wrong. She could not find the problem as she looked through the *rose-colored glasses* we mentioned above. She could see the problem as she was looking at the issue with the bias that she had done everything correctly as per the instructions. She pushed it over to me and asked if I would take a look. In less than a minute, I saw the problem. I pushed it back to her and told her to really look! I asked her a few questions.

One was: Throughout ALL the classes we have had in this program talking about electrical and electronic devices, are there any of them that are directional? Meaning they can only go one way! She looked again. Aha, she was able to see it now, she put the capacitor in backward! In a capacitor, one lead is positive and one lead is negative. If you put it in backward, it will explode! Well, it exploded! Luckily it was inside a metal case as it was a rather large capacitor.

What does this story have to do with creative thinking? If I were the instructor of an electronic course and I purposely had students solder in a capacitor in the wrong direction, then the student energized the device. Trust me: That student would always remember the BANG and the burnt electronic smell for the rest of their life where they would not let it happen again. Would this not be a creative way of thinking to get a concept or idea into the mind of a student? I rest my case.

Through all my years of being in management and supervisory positions, giving presentations, and being the previous owner of an IT corporation, I have seen ideas from many authors discussing and laying out what they think are traits of managers and supervisors. In addition to being a manager or supervisor, which is a leader in many aspects, they also need to add traits of a follower as they are not the top tier of the corporate structure; everyone has a boss, right? Below I will propose what I think to be additional hats relative to leaders and followers, which I feel run parallel to the college freshman for their goal in achieving more than simple success and passing, but for attaining above-and-beyond success. The success that needs to be achieved by the graduating senior in their attempt to compete in the marketplace with all the other graduates around the world. The layout I propose is to provide individuals a planned method to *think* cohesively to be more effective in attaining results. What does that mean?

The foundation of this layout is that the human brain goes through a number of instructions or steps, if you will, to develop a plan of attack to solve a problem. Each instruction or step is required and is a necessity to solve the problem or task successfully. The thought is that each instruction or step when activated will unconsciously bring forth certain aspects of thought for the other instruction or step to be triggered, therefore coming up with a viable solution.

I will say this. These *instructions or steps* are not ways of thinking that all college freshmen share or may even possess, but I do think, however, that these *instructions or steps* can be a learned trait. These instructions or steps can be attained through educational reinforcement and practice. Growing up, didn't we always hear *Practice makes perfect*? Well, here we are, practice does make perfect, at least we are told. The more you practice, the better you get at it, right? Is it not easier to review work that you know has been successfully completed, as in a template, than it is to start from scratch with no provision of an example? It is like asking someone to reinvent the wheel when you have access to a perfectly good example of a wheel, right? What is the point?

Do we need an example here? Okay, let me set the stage. The first post-secondary school I went to for a degree was a Flight college in the south. I was an Avionics major, and my roommate was a Flight major. My roommate had some experience in flying and he wanted to get more flight time and a degree. He was very particular about his aircraft and didn't leave many things to chance when he flew. It seemed as though he was always flying when other students were just kind of hanging out. One of the things he liked was that one of the flight instructors had a similar background as he had, and he also wanted to get a lot of flight time, so they generally scheduled their flights together, instructor and student.

In the realm of flight, there are two ratings pilots achieve: VFR and IFR. VFR is Visual Flight Rating, and IFR is Instrument Flight Rating. Both of these ratings deal with eyesight. Visual is what you see inside and outside the cockpit and how you deal with what you actually see visually and how you react and deal with it. Instrument is where you ONLY use the instruments to see and project what to do next in a flight situation. Make sense? So, my roommate had achieved both of these ratings. Remember, I said my roommate had spent a lot of time with one certain instructor during his time at the college. SO, on to my story.

I came back to the dorm one night, all the lights were out, and my roommate was sitting on the edge of his bed, very quiet. This was unlike him, so I knew something had happened that may be life-changing. Four hours ago, my roommate was up with a different instructor getting some extra flight time. He had to choose this other instructor as the one he usually scheduled with was up in the air with another student. My roommate and his new instructor are up in one airplane, and his usual instructor and another student are up in another. I don't remember what class-level storm came up, but it was rough, and both airplanes were caught in the storm. For over three hours, both of these airplanes were buzzing around the sky. Lightning struck my roommate's airplane, taking out all the internal and external lights. The only light still on were the instrument lights. The cockpit was dark, and you could not even see the one person sitting next to you.

The only time you could see something outside the windows was when the lightning flashed. The radio was damaged also, and every now and then, you could hear garbled talk from the other airplane and the flight base towers: In and out, in and out, bits and pieces of conversations. All my roommate could really hear was someone talking about 1000 feet. He didn't know whether it meant his aircraft was to go to 1000 feet or if the other airplane was to be at 1000 feet. He could not make out the whole conversation. The new instructor with my roommate was not at all comfortable with the situation, so my roommate took charge. Like I said, my roommate was a particular guy and had no problem taking over. He was confident in his training. If anyone knows anything about an airplane pilot, they are cocky and confident!

After several hours of struggle, it is pitch black outside, and lightning is flashing. My roommate hears an airplane engine, and it is close by. The lightning flashes, and right in front of my roommate's airplane is the other plane directly in front of him at the same altitude. My roommate said his whole life flashed before his eyes, and during the brief lightning strike, he could see the eyes of the others in the other airplane. He didn't hesitate, pulled up on the yoke, and he said it was like his aircraft was standing still, without motion. In that brief instant, it was over! All four were safe! The instructor of the other airplane pushed his yoke down.

The two students and instructors are alive today because my roommate was in charge of one airplane, and his usual instructor was in charge of the other. Those two had trained together many, many times for such a scenario as this.

My roommate is a highly successful pilot for a major airline today, and we talk all the time. Practice made this perfect.

How about another quick example, a little different? I tell you to run a race. I give you the starting point and the ending point, but I leave out any specifics or limitations of the race. Then YOUR mind begins to start sorting, calculating, and planning. Do I have enough information? Do I need to start at a certain time? Do I need to get to the endpoint at a certain time? What mode of transportation am I to use? How many other questions would or could a participant ask? We already can see how creativity ties in with the success of a student from the information previously presented. Creativity or Creative Thinking being one of the two main necessities mentioned in detail above along with Critical Thinking.

'Boonie' the Strategic Thinker

Why the Boonie in relation to Strategic Thinking? The Boonie hat or cap has a wide flexible brim that protects the wearer from harsh surrounding elements. Typically, you see the Boonie used by trained military personnel, personnel sent to carry out life or death missions using both creative and critical thinking. Hence, the strategic thinker uses both thinking ideologies.

The strategic thinker is split into four distinctive groups, each one which guides to the next group, to a final outcome. They are: understanding a strategy; analyzing the position; planning a strategy; and implementing the strategy.

The first section called Understanding a Strategy, speaks of the first basic components that you should consider when looking to define a strategy and the basic components involved in thinking of solutions in a strategic way. Are you as an IT manager looking for short-term results or are you looking for future results? The IT manager should look at the employees and staff involved and seek to find which is better suited to assist in the process to find a solution which will achieve success. The IT manager can take a lot of waste and lost time out of the equation if you avoid guesswork and stick with the facts and statistical data. For success in the beginning stages and even in the final stages you must always keep reviewing the process to allow you to stay on track and not lose sight of the purpose of your project.

The second section called Analyzing the Position, speaks basically of knowing the audience you are getting an answer for, knowing what influences the results, knowing your audience needs, knowing your competitors, assessing yourself and your IT employee's skill and abilities and summarizing the data analysis of all of these factors combined. Every problem has at least one influencing factor this could be anything from monetary values to professional notoriety. What exactly is required of you by the company who has hired your IT firm or is the issue in-house? What are the audience expectations and limits? What are the aspects of competitors? What skills and knowledge does your IT solution team possess? When you combine all of this information together you will be better able to come up with selecting a plan of strategy that will assist your IT business in achieving success.

The third section called Planning a Strategy, speaks of the five stages to consider when completing a strategic planning project. Define the purpose, determine the advantages, set the boundaries of the project, choose the main areas to emphasize or which stand out, and lastly estimate the amount of money and time which will be allotted for the project solution. The IT manager should make sure that everyone involved in this project and solution is on the same page so that there is no conflict in either the project team camp or the customer's camp. Toward the end of this section there are two final steps in this process, to test to validate the strategies potential success that you have compiled and to make sure that communication has been successful and informed all of the people who need to be involved in this project.

The fourth and final section called Implementing Strategy is I would say the hardest step to complete. This step involves every single person at every level. When implementing a strategy there are several steps in the beginning of implementation which are setting priorities for the change: actually planning the change; assessing the potential risks involved in the change; and finally reviewing the targeted goals presented. Now we come to the hardest part that of motivating the staff. The majority of people I believe are scared of changes. Any internal or external movement in a corporate setting usually puts all employees on pins and needles. Usually if there is a change in the corporate situation that means that someone has a bright idea of improving something and that generally means that the improvement is going to benefit the company and not the employees.

You can bet that if the strategy benefits the shareholders that it matters not if you are a part of it or not so the only choice you have is to go along with it and hope for the best. After the motivational step there are three final steps to go through and they are monitoring the performance of the project after implementing, reviewing and analyzing the acquired data, and being flexible by looking at the data and if there are changes or alterations to be made to be successful then you must be open to suggestions and maybe even starting this process over again.

8 ... U P D A T E S ,
U P G R A D E S   O R
P A T C H E S

There are a few terms that are confusing and often mistakenly used as equals within the context of security. These terms are updates, upgrades and patches. Updates enhance existing software, upgrades involve major changes and or new versions, and patches provide urgent fixes.

If I were to give a classification to each of these as to their level of necessity or severity I would start with level one as the least important and level three as the most important. Not to say and of them are any less important, they each have their own usage.

U p d a t e s

I see updates as level one. Updates are generally improvements to existing software that add new features, improve performance and or augment security, and are released sometimes monthly or quarterly. When you buy a virus protection software package you generally get a number of free updates after the initial purchase. These updates are where you would receive a file and install that file into your existing package then be supposedly safer than the last update you previously installed.

Have you ever been to Graceland in Memphis Tennessee, the former estate of Elvis Presley? This estate was decorated during the 1960s and 1970s previous to the death of Elvis. If you were to buy that estate and update the décor then you would replace the existing shag carpet, dark colored drapes, and the extreme bright wall paint colors. You do not have to do it, you just want to do it. I guess the key work here is "want" and not necessarily a need or necessity.

If you relate this to a virus software program, you may be fine by not updating the software, but you may put yourself and your systems at risk by not updating.

U p g r a d e s

I see upgrades as level two. Upgrades generally offer major changes and or improvements, and possibly redesigns, and usually released less frequently than updates. If you have the same virus software package and after a certain amount of time you may no longer receive free updates and they offer you an upgrade to a newer version. This newer version could have any number of reasons for its purchase over the version you already have.

This upgrade may not necessarily be needed currently, but you may be projecting for some future purpose or foreseen need. If this upgrade is purchased generally you will uninstall the previous version and install the newer version in its place. I see an upgrade as a "need" not a "want" as in an update.

Maybe the program you use currently has all the features you need and use and you have no use in paying the money to buy the upgrade. This happens quite often. Some programs do not let you skip upgrades. I remember years ago when purchasing Auto Desk AutoCAD software our instructors had no real need to purchase new versions as the versions they used in classes had all the necessary features. The problem was, when it finally became time to upgrade to a version they thought was beneficial, they would have to purchase all the previous versions they skipped to be current. They could not save money by skipping versions.

Patches

I see patches as level three. Patches I see as the highest level three as they are critical. Patches are released at no specific time intervals to fix vulnerabilities found within software, to address urgent issues and or bugs, and need to be installed to prevent possible failures. It is crucial to implement patches as soon as they become available. Patches become available when companies are made aware of issues of holes that exist within their current package. Possibly people within the company find these issues and or customers and user outside the company find them and issue complaints or notices.

Patches can be simple or complex. They can be installed on a small or large scale. Larger companies usually have test and production servers. You would install the patch on the test server first and see if any issues arise within an allotted amount of time, then if there are no issues then installation on the production server would take place.

9 ... C R I M E

Several questions come to mind when we think about crime or situations involving some type of crime or element of criminality. Is that crime civil or criminal? Is it either a misdemeanor of a felony?

C i v i l   o r   C r i m i n a l

Determining whether an act is classified as a civil or criminal offense depends on several factors, including the nature of the act, the intent behind it, and the legal framework of the jurisdiction in which it occurs.

C i v i l

Civil offenses (also known as civil wrongs or torts) are wrongful acts that harm or injure another person or entity's rights or property. The intent of civil offenses may involve intentional, negligent, or strict liability conduct that causes harm. Penalties for civil offenses typically involve monetary compensation (damages) paid to the injured party to compensate for losses. Civil offenses are resolved through civil lawsuits, where the injured party (plaintiff) seeks remedies such as compensation or injunctions against the responsible party (defendant).
Examples of civil offenses include negligence, defamation, breach of contract, and certain types of cyber incidents that result in financial losses or privacy violations.

Criminal

Criminal offenses are actions or behaviors that violate criminal laws established by governments to protect public welfare and safety. The intent of criminal offenses typically involve intentional or reckless conduct that harms others or society as a whole. Penalties for criminal offenses may include imprisonment, fines, probation, or community service. Criminal offenses are prosecuted by governmental authorities (e.g., district attorneys, prosecutors) on behalf of society.

Examples of criminal offenses include murder, theft, assault, drug trafficking, and cybercrime such as hacking, identity theft, or cyber terrorism.

Differentiation of Each

Some actions may overlap and be both civil and criminal offenses. For instance, a cyber-attack that results in financial harm (civil) and also violates criminal laws against hacking or fraud (criminal). The burden of proof is higher in criminal cases (beyond a reasonable doubt) compared to civil cases (preponderance of the evidence). The objectives of criminal law are punishment, deterrence, and rehabilitation, while civil law focuses on compensation and prevention of future harm.

In the context of cyber terrorism or serious cybercrimes, such as those intended to cause harm to national security or public safety, these are typically classified as criminal offenses due to their severe impact and intentional nature. Civil remedies may also apply, especially in cases where individuals or organizations seek compensation for damages resulting from cyber-attacks. Therefore, whether an act is classified as a crime or a civil offense depends on the specific circumstances, legal interpretations, and applicable laws in each jurisdiction.

Misdemeanor or Felony

Determining whether a crime is classified as a misdemeanor or felony depends on the severity of the offense and the specific laws of the jurisdiction in which the crime occurred.

Misdemeanor

A Misdemeanor is a less serious crimes like petty theft, disorderly conduct, simple assault, etc., with less severe penalties usually involving fines or imprisonment for less than one year. Whereas, a felony is a more serious crime such as murder, rape, robbery, arson, etc., typically punishable by imprisonment for more than one year.

Misdemeanors are less serious crimes compared to felonies. They typically involve minor offenses that may not result in significant harm to individuals or society. Penalties for misdemeanors usually include fines, probation, community service, and imprisonment for up to one year in local or county jails. Examples of misdemeanors include petty theft, simple assault, disorderly conduct, and certain types of fraud involving smaller amounts of money.

Felony

Felonies are more serious crimes that often involve significant harm to individuals, property, or society. They are typically punishable by imprisonment for one year or more in state or federal prisons. Penalties for felonies may include substantial fines, lengthy imprisonment (typically more than one year), probation, and other consequences that can affect civil rights. Examples of felonies include murder, rape, arson, burglary, aggravated assault, drug trafficking, and serious cybercrimes such as large-scale hacking, identity theft, or cyber terrorism.

As They Relate to Cyber Crime

Cybercrimes that involve significant financial losses, harm to critical infrastructure, or threats to national security are often classified as felonies due to their serious nature and potential impact. The classification of cybercrimes as misdemeanors or felonies depends on the specific cybercrime laws enacted in each jurisdiction and how they categorize offenses based on severity, intent, and consequences. Laws regarding cybercrimes can vary widely between countries and even between states or provinces within a country, leading to different classifications and penalties for similar offenses.

In summary, whether a cybercrime such as cyber terrorism is classified as a misdemeanor or felony depends on the specifics of the offense, the jurisdiction where it occurred, and the legal interpretations of applicable laws. Given the potential severity and impact of cyber terrorism, it is more likely to be classified as a felony due to its intent to cause significant harm or disruption to national security or public safety.

10 ... THIS COURSE: BLACK HAT PEN TESTING

Out with the old and in with the new. We all learn from past experiences.

What makes this course is unique is several items. The first item is there are no guidelines, rules, regulations, or ethics involved in this course. There is no red team/blue team where there is a delineated offense and or defense. There is no one watching your connectivity nor tracking your entrances and exits in trying to breach these three victim stations. These three victim stations and the predators you will be using are all on a separate network from any and all campus devices. Other than the research and preparation portions required in this course all predator to victim access will be a 100% requirement of the physical network classroom/lab. This is not a virtual classroom experience. Roll your sleeves up and get ready for the most comprehensive course you will have taken at this point in your university career.

Scenario

You have just graduated and a company has taken you on as a new employee. This company takes it for granted you know something of cybersecurity as you just graduated with a BS degree with honors. Are they correct in this assumption? Did you pay attention in your degree studies or just go through the motions to do just enough to get by and pass? My guess is that you graduated with honors so your grades must have been relatively high in most courses.

The Task

Your boss is interested in penetration testing. He wants to develop a step-by-step course for all future employees to take making them aware of cybersecurity flaws and fixes within working environments. He has decided to hire a team of graduating students with expertise in the cyber and security realm of expertise. At the end of the task period he proposes to have a professionally complete portfolio.

If you were a full time employee handing this task you would be working on it for 8 hours a day five days a week for 15 days until an expected completion. This will not be the case here. In an educational environment you will be given 16 weeks to complete the task. This will be a proposed 8 hours per week for 16 weeks until a necessary completion. The hours taken will be up to each student participant. Trust me when I say this: It does not matter how well you think you have a friend that will let you ride on his/her coattails to copy/paste their work. If they devote the necessary time to complete and pass this course they will most certainly not give you their work.

The task is in the allotted time to complete a comprehensive portfolio and the allotted assignments.

Software

No software in this course will be purchased. All software used by the student will be free. This freedom of thinking will give the student an edge in their journey of "thinking like a hacker". Hackers do not buy software. The thinking behind a successful hacker is to not be seen, traced or tracked and to hide behind hardware and or software and to give the illusion you are not who you are nor your intent. This can only be carried out through the attainment of free untraceable and non-serialized software.

Textbooks

The first textbook chosen for this course is titled "Penetration Testing Fundamentals". What better way to teach network personnel of offense network strategies than to use the structured thoughts and concepts used by Penetration Testers, often mentioned as Ethical Hackers? This textbook starts with the simple ideals and concepts behind some of the thinking of the potential hacker and goes into more depth as the chapter's progress.

Penetration Testing Fundamentals: A Hands-On Guide to Reliable Security Audits
Author: Chuck Easttom II
Softback: 978-0-7897-5937-5

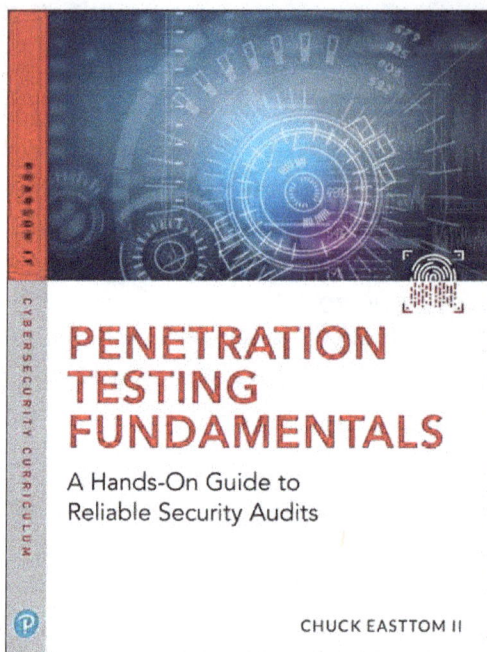

Generic Image: Penetration Testing Fundamentals

The second book you will need is the one you are reading at this time. It is a pdf and you will be given the digital version. If you are in this course you are reading it now.

Cybersecurity
Black Hat Pen Testing: Post-Secondary Offensive Exploitation and Breaching
Author: Clifton Ray Wise
Softback: 979-8-989923-69-4

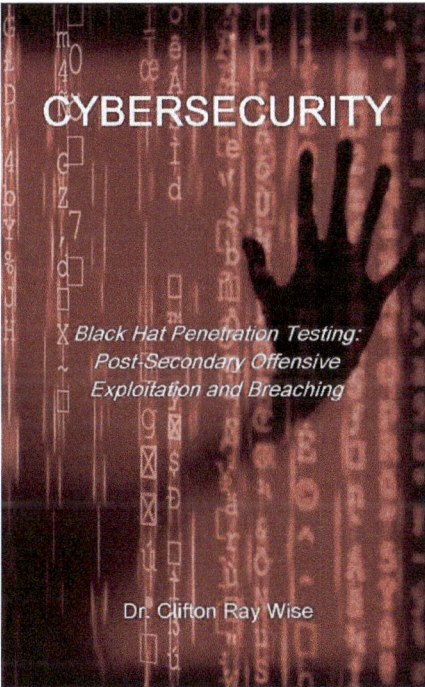

Generic Image: Cybersecurity Black Hat Pen Testing

The third textbook chosen for this course is titled "Web Penetration Testing with Kali Linux". This textbook starts with the basics of the Linux operating system as well as some navigation and concepts behind some of the thinking of the potential hacker and goes into more depth within Linux as the chapter's progress.

Web Penetration Testing with Kali Linux
Authors: Joseph Muniz and Aamir Lakhani
Softback: 978-1-78216-316-9

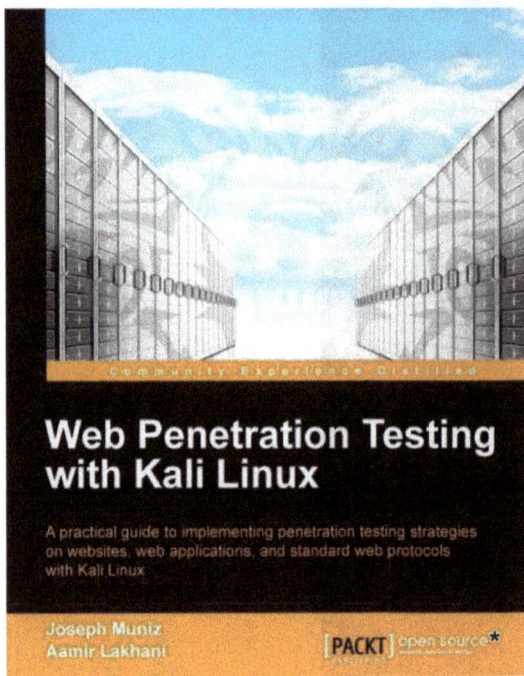

Generic Image: Web Penetration Testing with Kali Linux

A b o v e - a n d - B e y o n d

In case you are one of those students who go above-and-beyond just doing the bare minimum to complete a course? In addition to the three texts listed above you can do some brief research on the web and find additional helpful useful materials for this course. Some of these are listed here in addition to others you may find:

- The Hacker Playbook 3: Practical Guide to Penetration Testing
- Common Ports.pdf
- Wireshark Display Filters.pdf

- 20 Fantastic Kali Linux Tools
- CEH Certified Ethical Hacker: Exam Guide
- Common Kali Linux Commands Overview
- Kali Linux Cookbook
- Penetration Testing Fundamentals: Coalfire
- Mastering Kali Linux for Advanced Penetration Testing
- Why Should I Care? Mobile Security for the Rest of Us
- Wireshark Quick Start Guide

Physical Connectivity

ALL physical interaction between the Predator laptop and the victims MUST be physically in the lab HH328. There is NO other way to reach the victims! They are NOT on the KSU campus network!

There are 3 victims: Windows Workstation (un-patched and non-updated with a static IP), Windows Server (un-patched and non-updated with a static IP), and a Linux box (un-patched and non-updated with a static IP).

I find it still necessary to complete this course physically.

You will use one of the lab workstations to carry out your research and documentation for this course. ANY of the workstations in the lab/classroom.

You will need to have a USB flash drive. This is where you will put any software you download for this course. You will the transfer those software programs to the Predator laptop for installation and or running. ** *These* *software's are only to be installed and or ran on the Predator laptops.*

You will be given a physical laptop (Predator). This Predator laptop will have access to the Victim switch where the 3 Victim boxes are connected. ***Remember the number of laptop you have, at the end of each days use put the laptop back in the cabinet and plug up to change. This will be the Predator you will use during the duration of this course.* When booting up the Predator you will have two choices: Windows or Linux? Choose the one you are working on.

You will use a straight-through patch cable to plug your Predator into the BLACK port behind each/any workstation you have chosen in the lab/classroom. This will be your physical connection to the server rack patch panel in the back of the network lab/classroom HH 328.

You will go back to the network rack patch panel (BLACK) and use a straight-through patch cable to plug the Black Port number of your station into the Victim switch.

Predator (Laptop) – Port (Black) – Patch Panel (Black) – Switch – Victims (1, 2, and 3)

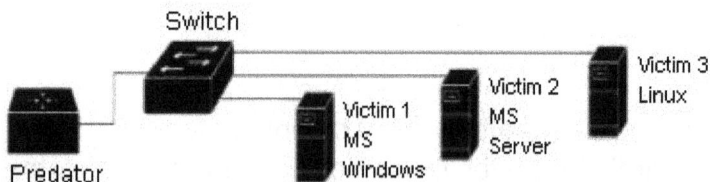

Switch

Predator

Victim 1
MS
Windows

Victim 2
MS
Server

Victim 3
Linux

Victim Stations

These are the 3 victim stations: They will each have a static IP that will not change during the duration of this course. I would still go though some of the work and research to see how you find IP addresses and computer names. In the real world some IP's will be static and some will be dynamic. ***I would suggest "pinging" the Victim you propose to work with at the beginning of each work day to make sure that Victim is up and running and has not been turned off for some reason, over the weekend, vacation, etc.*

- **Victim 1: MS Windows PC**

IP: **192.168.0.5**

NAME: **SAM**

USER: _____

PASS: _____

- **Victim 2: MS Windows Server**

IP: **192.168. 0.10**

NAME: **X-Ray**

USER: _____

PASS: _____

- **Victim 3: Linux**

IP: **192.168.0.15**

NAME: **Lima**

USER: _____

PASS: _____

***Predator Laptop Login Credentials

USER: **csguest**

PASS: **P@$$4csg**

**Every operating system you are proposing to breach has a date of manufacture and dates of patches and or updates. These are readily available on operating system websites. If you realize that the OS install has not been patched or upgraded then you can easily research possible holes and or open ports.

D e l i v e r a b l e s

Along with the delineated weekly assignments within canvas for this course.

YOU will turn in your completed portfolio at Weeks 4, 8, 12, and 16. There will be NO extended deadlines! Each of those 4 submissions will be graded. This is NOT the class to scrimp on time spent and procrastinate. This is NOT the class you will do the minimum amount of work just to get by and get a grade of "C". The minimum grade of a "C" just may end up being a "D". All students in this course are expected to get an "A".

This is a high level course. It will take time. After a few weeks you will come to some realizations. Things will start to click. The left side of your brain will start to pulse with the right side of your brain. Thoughts will start coming together. Things will start to make sense. You will find yourself wanting to spend more time reading, thinking, planning, and attacking!

P r o p o s e d   P o r t f o l i o

Cover Page

Daily Log of what you do related to this course

Example:

6-15-24:

Read Syllabus, Read book 1.

6-18-24:

Read Chapter 1 of book 2.

Completed end of chapter questions and activities.

Completed Week 1 Canvas assignments.

Downloaded programs discussed in Chapter 1 ___named here___, transferred to Predator, installed and ran.

From the program ___named here___, this is what I found. I found the program to be helpful/not helpful.

This is what I found

_____

_____

_____.

6-24-24:

""Any time you spend on this course is to be thoroughly documented. You do NOT need to document the specific amount of time spent. You will be graded on this portfolio.

11 ... MY PREVIOUS COURSES: OFFENSIVE NETWORK STRATEGIES

This course was previously offered for undergraduate graduating seniors and graduate degree students. I offered one of the classes before the covid epidemic and one after the covid epidemic. At that time I was not looking at any relationship to the students and their prowess related to the epidemic it just turned out that is when each course was offered.

You could say this was a controlled environment and the only two changes made between the two courses were, there were only undergraduate students in the post-covid course and I changed the victim stations to having static instead of dynamic IP addresses, this saved a lot of work for the participating students.

What information was I seeking in this experiment? I have always been an advocate for realism. Realistic environments. I have spent the majority of my life in industry dealing with vendors, customers, supervisors, employees, etc. Dealing with the whole gambit of relationships within industrial situations. In reality you may or may not have specific training. In reality it is expected if you have a job or career position that you know what it takes to handle that positions and to do it to the best of your ability, or you will get replaced. It is that simple. You rely on critical, creative, and strategic thinking to assist you to keep your job. There are no rubrics to follow to give you an idea what exactly you have to do to pass and keep your position. Nobody sits down with you and trains you in the position you were hired for. Unless that is part of your package. It generally is not. There is no one there to hold your hand. You might get a first-week of leniency but after that it is up to you. The key I am pushing here is: It is all up to you!

Pass or fail it is all up to you. It is up to you to ask questions if you have no answer. In the real world you ask someone around you who may have the answer. Passing or failing is up to each student. It isn't a group effort as you will pass or fail as a team. You are your guide and conductor. It is your responsibility to pass or fail in life, in your profession, and in your education. Why not make some of your education realistic and not exactly as what they book says. If you go into any industry there are tips and tricks and they are not in any book.

Implementation Results

Pre-Covid

The pre-covid implementation went as expected. Each of the students got together and worked as a group majority of the time. If those students had questions they asked those questions to me as I was the instructor. In the past I had all of these students in multiple courses I had taught. There had never been any issues with students catching me anywhere I was to ask questions about this course. This course went off without a hitch as normal. I covered all the information on the first day of the course. Few students took any forms of notes but they were all present and they all paid attention as this was their last semester and they wanted to graduate. One of these who graduated, a master degree student, now teaches at a community college a pen testing course and does it my way, the realism way. He says it is very popular.

P o s t - C o v i d

The post-covid implementation was the total opposite from the first implementation. I covered all of the material the first day of the course. Two students were late, one of them was late to most every class of mine the last three years anyway.

Once all the participants were present I went through the materials and escorted them around the room showing them all the particulars. There were no note-takers but this was normal as I generally have 2 note-takers a year and that about covers all my courses. I proposed the course meet as a team when the students wanted or needed to meet as a team. This course met once a week and was at the discretion of the students. Now, I see all these students in other classes I have and at other locations throughout campus. I see them all the time. Sometimes, I grab them and ask how things are going in this course. I never get any questions.

In departmental meetings instructors inform me my students are asking them questions about my course. So I make the decision we need to meet during the prescribed course time to go over questions if there are any. This is the end of the second week and no students have asked my anything.

I meet with the students. The same student is late as usual. I go through the same structure and routine I went through the first day of class. My first question: Did anyone take notes the first day of class? The answer is no. At this point, I would expect students pick up a pen or pencil and start writing. Well, the answer was no again, no note takers this time either. None were interested in taking notes.

I asked if any of them had started reading the guide for the course as it has a step-by-step guide of how to proceed in the course. The answer was a resounding no, they did not know what I was talking about and where was it. It was very confusing. None of these students cared to take any notes. They blamed me for not hand-feeding them in this course. It is 100% my fault they do not know where or how to start.

Now, these are all graduating seniors and this is their last semester. None of them took notes for anything. One of them asked for a rubric to be handed out so she made sure she knew what exactly she had to complete for points to pass the course and graduate. The students who was always late even pointed the finger at me the instructor for all the confusion. He mentioned something about being thrown to the wolves.

So, I knew at this point none of those students had any idea how to complete this course on their own. I had to step back and make typical quizzes, a midterm and final multiple-choice exam, as well as other creativity and critical thinking lacking activities for them to pass the class.

At the end of the course two students mentioned I had thrown them to the wolves and the others said it was one of the best course they had ever had as they had never had a course like that and they would have liked to have had more like it.

Pre-Covid Implementation

The premise of this project is to document the implementation of an Offensive Network Strategies themed special topics course for undergraduate and graduate levels, i.e. COS 499: Special Topics in Computer Science and COS 599: Special Topics in Technology.

COS 499: Special Topics in Computer Science. This course may have a different topic relevant to the computer science field each time it is offered. The purpose of the course is to gain knowledge in current areas of this ever-changing field. The course may be repeated four times for credit. Prerequisite: permission of the instructor. Credit: 3 semester hours.

COS 599: Special Topics in Technology. A treatment of topics relevant to the computer security, networking, or information technology not routinely covered by other courses. New developments in systems concepts, techniques, and equipment. May be repeated up; to 2 times for credit. Prerequisite: Permission of the instructor. Credit: 3 semester hours.

In the realm of the growing topic of Cyber Security today, there exists human entities which daily threaten our public and private networks. Within the field the term for these entities are Black Hat and Gray Hat Hackers. The premise of the Black Hat is to attack networks with negative intent, whereas, the Gray Hat is to attack networks with possible positive thoughts, all leading to the same conclusion, exploiting vulnerabilities with criminal intent.

Within the educational jurisdiction for years the idea was to solely teach courses in defensive network strategies. Strategies to use generally purchased software to help the student to understand how to defend a network. To successfully defend a network the student must grasp topics from architectures to tracking. Solely understanding defense techniques of a network, however, necessary have driven educators with the concept of adding offense network strategies to the mix.

Within the field of Cyber Security we know that a network is only as secure as the money and knowledge we attain to keep it safe. Security is not something we just set-and-forget. Successful security is implemented through constant training, patching and updating of technologies. "Keeping current" is a term often associated with this field of expertise. The team that does not keep current is the team that often fails or becomes a target of exploitation.

What better way to keep current than to train network personnel of Offensive Network as well as Defensive Network Strategies. To not only treat the student as a potential White Hat hacker, but to also give them the tools used by the Black Hat and Gray Hat Hacker as well. To understand the mindset and concepts thought and learned by the potential network harasser and attacker, by using hierarchies and software's generally used by these human entities, therefore, the premise of this course.

Post - Covid Implementation

This was the most comprehensive course any of my students have ever taken. The comprehension is up to you.

There are a few defensive strategies we talk about but mainly this courser is about offensive network strategies.

Think of a football team. Half the team is defensive and half the team is offensive. The defense defends the offense from attacking and making a goal and the offense tries to make the goal.

What is the goal? The goal is different depending upon who is trying to make the goal (Money, Fame, etc.).

The typical defense is simple: Keeping patches and updates current. There may be a person or team watching network traffic come in and out of a network. There may NOT be anyone sitting in a chair watching, but if the network administrator setup the network correctly he/she will be warned of possible intrusions.

The typical offense is more complex, and takes more time. Is someone watching? This is what makes it more complex. Is someone watching? How are they tracking if they are? How much time do you have to try something before an alarm is set? Where can you try something? What can you try? ONE false move or WRONG step and you are FOUND and in jail! In jail with a monetary penalty and lawyer fees! IS it worth the risk? How big is the risk?

A hacker just doesn't wake up one morning and say he is going to try and hack into something! If he or she does? He or she is most likely in jail!

A BIG portion of the time a hacker spends is on research. An investigator of sorts. This is what he or she spends the majority of time on. Researching and investigating the victim. If it is a network then the victim is likely ONE device. ONE router, ONE switch, ONE firewall, ONE laptop, ONE desktop, ONE server, etc.

Each device has a different way to break into it. Different operating systems (OS) have different ways and different programs to retrieve information from them. ONE size does NOT fit ALL!

TWENTY FIVE years ago there were not that many computers in the world, not like today anyway. The internet was in its infancy. NOT everyone had access to computers and or networks and or the internet and Google!

YOU can find out a lot of information about a victim on the internet. Sometimes "victims" have a BIG mouth! Sometimes they talk TOO much! Sometimes they give out secrets and don't even realize. Research and Investigate. Find out about your victim.

What types of networks do they have? Are they internets or intranets? What operating systems (OS) do they use? Have devices been patched or updated? When was the last time they were? What was in the latest patches they may not have installed? How do I know MY victim is: up, on, running? I can't hack a device that is off!

PING is a simple command. Can someone see me if I PING their system?

Stealth is the key! Research and Investigate your victim prior to an attack! Research and Investigate software? Trial and error! What do they tell you? Are they useful? Which are the useful ones? What do others say?

A P a r a l l e l

I spent quite a few years within what was once called the Punch-Kick-Association, better known as the PKA. I would guess most people think 'Karate' is the same everywhere and is the all-encompassing word to describe all things kicking and punching. That is pretty much a misnomer, there are many types or styles you might say and they come from different locations and countries around the world. My style was Taekwondo, which originated in Korea, so you would say it is a Korean martial art. Anyway, I mention this as when you attend a 'hacking' competition I see quite a few parallels when I look at both competitions. When you first walk into the venue it looks like random chaos, but in fact it is quite controlled.

Generally you will see a lot of groups spread out. Each group is separated for one reason or another. Each group has a referee that pays attention to all aspects of each group. Sometimes there are bouts, sometimes there are shows, etc. There are a lot of things happening at the same time.

Why do I mention this? Well, if to kind of give a preview of a competition. A competition related to cybersecurity, red team – blue team, whatever it is called. Generally, they are not quiet places to think and you need to prepare yourself to block out the sounds and commotion around you. This is where it is imperative you know what to do, when to do it, and the order in which to carry it out. Sometimes these competitions can be like a casino. If you have ever been to a casino? All the noises, the bells and whistles, the flashing lights, etc. Although, casinos do this for a purpose: To keep the players and attendees confused.

There is a certain amount of respect shown at events. At a Karate tournament, if you are slightly versed in the etiquette of those you know: When you come into the doorway at an event, you stop, you put your hands to your side, and you bow from the waist when entering. This shows you are giving respect to the attendees, the participants, the referees, and the sensei's (teachers). You give a similar form of respect at other competitions by listening, not being disruptive, and following along with the program.

In Preparation

My advice is to acquire a:

- notebook about one inch thick
- spiral notebooks with holes and tear out pages if you like those (college ruled)
- good writing pen or pencil

You will be keeping a daily portfolio. You can review the syllabus and decide whether to follow the weekly plan or make your own plan. As long as all the material is covered and in the portfolio, I do not care how you proceed, it is up to you and your team.

In This Course

Dependent upon the number of students in this course you may be split into teams. If split into teams, each team will work together to present team portfolios the end of weeks 4, 8, 12 and 16. The end of week 16 all teams will work together and submit a single portfolio using the best from portfolio portions from all teams. You will be given a Team Leader by the instructor. This Team Leader will be responsible for getting participatory effort from all members of the team and providing grade input for each member of their team at midterm time and final exam time.

YOU will start and keep a portfolio. Every day you spend time working on this course you will document that time.

There will be THREE books used in this course. The Pen Testing Fundamentals is the go-to book you will be using. This will be YOUR Bible in this course. The second book is about Cali Linux. It will be a reference but will be VERY helpful when you start the Linux portion of this course. The third book will be this book which will give you details and specifications of this course.

Other books may also be made available.

At the end of each chapter of the Pen Testing book there are Questions and Activities. The answers to the questions and activities will also be in your portfolio.

ALL this course revolves around YOUR comprehensive portfolio!

YOU will turn in your completed portfolio at Weeks 4, 8, 12, and 16. There will be NO extended deadlines! Each of those 4 submissions will be graded. This is NOT the class to scrimp on time spent and procrastinate. This is NOT the class you will do the minimum amount of work just to get by and get a grade of "C". The minimum grade of a "C" just may end up being a "D". All students in this course are expected to get an "A".

This is a high level course. It will take time. After a few weeks you will come to some realizations. Things will start to click. The left side of your brain will start to pulse with the right side of your brain. Thoughts will start coming together. Things will start to make sense. You will find yourself wanting to spend more time reading, thinking, planning, and attacking!

For the Specifics

There are TWO ways of proceeding through this course. The previous times I have taught this course it was a physical course and all interactions between the predator and the victims was with a physical hard wired connection. I have proposed a new way through remote access. I am not sure if this will be up and running and available in future semesters?

Physical Connectivity

ALL physical interaction between the Predator laptop and the victims MUST be physically in the lab HH328. There is NO other way to reach the victims! They are NOT on the KSU campus network!

There are 3 victims: Windows Workstation (un-patched and non-updated with a static IP), Windows Server (un-patched and non-updated with a static IP), and a Linux box (un-patched and non-updated with a static IP).

YOUR ultimate task is to BREAK into one of those boxes! To BREAK into those boxes by going step-by-step through the Pen Testing book.

The 3 victims are un-patched (plugged into a switch).

Each DESK in lab HH328 is plugged into the victim switch. (What color?)

You will be given a dual-boot laptop (Predator). Put a sticky note on the ONE you will be using as it goes back in the cabinet when you are done for the day. Plug it back into power in the cabinet. This LAPTOP will be the ONLY device YOU will use as the Predator! NO OTHERS! Your dual-boot predator laptop will be with MS Windows and Kali Linux.

That laptop will have a patch cable. That patch cable will plug into the port listed above (What color?)

This laptop will be physically and directly connected to the victim switch above.

These are the 3 victim stations: They will each have a static IP. When this course was previously taught, the victim stations were given dynamic IP address, and they changed quite a bit during the course. This time you will be given static IP addresses that do not change. I would still go though some of the work and research to see how you find IP addresses and computer names. In the real world some IP's will be static and some will be dynamic.

- **Victim 1: MS Windows PC**
IP: **192.168.0.5**

NAME: **SAM**

USER: _____

PASS: _____

- **Victim 2: MS Windows Server**
IP: **192.168. 0.10**

NAME: **X-Ray**

USER: _____

PASS: _____

- **Victim 3: Linux**

IP: **192.168.0.15**

NAME: **Lima**

USER: _____

PASS: _____

***Predator Laptop Login Credentials

USER: **csguest**

PASS: **P@$$4csg**

R e m o t e ' F u t u r e ' (i f
a v a i l a b l e)

This will be using the back door of our Cyber Security Server. There will be three (3) victims setup using Virtual Machines (VM). These victims will use the same operating systems (OS) as denoted as in the physical connectivity. These victim stations will each be given static IP addresses.

There are two different ways to proceed with the predator stations, either also providing them as virtual machines (VM) or providing them as the laptops provided in the physical connectivity. At this time I have not decided which would be the best solution for learning. As noted above, remote access is not available at this time.

If you are given a VM to use as a predator you will be given a link to your station. You will be given access information similar to this:

- **Predator VM: Windows**

IP: 192.168.90.18

NAME: _____

USER: _____

PASS: _____

You may also be given a separate VM for use with the Linux portion of this course. If you are that information may be similar to this:

- **Predator VM: Kali Linux**

IP: 192.168.90.18

NAME: _____

USER: _____

PASS: _____

T h e   P l a n

You will have access to the desktop workstation. This will be used to research your victim. To download software. To access the internet. *****YOU must NOT use this desktop workstation to install any of the software you will download! If you download software you will transfer that software to a flash drive and then transfer it to the Predator laptop. There YOU can install any software you like in your research and or attack!

My Suggestion

Read the Pen Testing book FIRST! This will give YOU a real good idea of what you are about to do! You will get some ideas this way!

*The portfolio will be hand written. By physically hand writing you add one of the human senses (feeling). Seeing is 20%, Hearing is 20%, and Feeling is 20%, for a total of 60%. (You cannot Smell or Taste so that 40% is not available). YOU will be surprised by how much more knowledge you can attain through the simple act of writing. Same with taking notes. Note takers learn more!

Structure your portfolio anyway you like!

You can review the weekly structure I proposed (windows to Cali- Linux) in the syllabus OR you can go chapter-by-chapter in the pen testing book, up to you.

My Goal

I physically know the author of the Penetration Testing book. My plan is to send him a copy of the final portfolio of this course and have him review it and possibly have him come up with a newer edition of his Penetration Testing book.

Necessity Warrants Research

For four years our Historically Black University (HBCU) has mismanaged approximately 36.5 million dollars. Our university has recently been overcome with breaking news articles surrounding monetary issues, a resigned president, and a new presidential search. Our computer science department is known for its aggressive presentation of proposals. Proposals to aid in the transformation of our limited classroom technologies to teach our students so they can learn cutting edge technologies as well as progressive thinking. During this term of years we have gotten approval for several proposals, although, the money has become an issue.

This monetary issue has lead us, the computer science faculty, to seek other means of trying to keep our students on the cutting edge as best we can. This seeking leads us towards thinking creatively. One critical question: Do we necessarily need money for top-of-the-line equipment to teach top-of-the-line cutting edge technologies? After reviewing the question at hand, the answer is, not necessarily. This first question leads us to a second question: Are we already in possession of equipment that can be used to teach what we consider to be cutting-edge, offensive network strategies?

After review, the answer is yes, we do have equipment within our area we can use for our projected implementation. Since we have equipment for our coursework, this brings us to a final choice. Our choice is to either combine material and a structure for our students or we need to choose what we feel is an already produced guide for student success, hence, leading to the chosen textbook for this implementation.

Past Implementation

This course was previously taught as a COS 499/COS 599 Special Topics course. This course was a mix of undergraduate and graduate students. This first deployment of this course there were the same three victims, although the course was a physical on-campus course with no reaching of victims from off-campus. The students were to be on campus to have access, therefore, the future change to off-campus access through the move to a specified VLAN. A second change from the past to the future is the upgrade and implementation of a static IP address for the victim stations. The third progression was to move from each student receiving a laptop as a penetrating attacker to each student using their own devices, also giving the necessity of using the specified VLAN mentioned above.

Pen Testing or Offensive Network Strategies

Pen Testing is the ethical training and implementation of software to find holes within a network or networks which can be exploited by potential hackers for the general purpose of data corruption or criminality. Pen Testers generally trained through the use of costly prepackaged software tools and a hierarchy of processes to gain potential entrance. The Pen Tester plays the role of an offender through ethical means. The Pen Tester would have previously had some coursework or training relative to the basic defense of networks and or computers. The opinion of this writer is that in the world of today there may be more use in training Pen Testers via the use of the thinking and methodologies behind the offensive and defensive strategies hackers use in real life. When a Pen Tester is engaged by a corporate entity a several of the beginning stages which would normally by thought out and addressed by the hacker are presented to the Pen Tester through documentation and nomenclature, IP addresses, OS brands, etc. The hacker does the research and puts together the documentation. The documentation and processes he/she will use to exploit their victim. The premise behind the guise of the hacker is stealth. The hacker preys upon their victims by the use of free untraceable tools shared throughout the World Wide Web. The bottom line is the corporate entity has hired a Pen Tester to ethically find holes and the Offensive Network Strategist is knocking on the door of your network trying to gain unwelcome access.

H a r d w a r e

This course will offer three computers to be setup as "Victims" with older un-patched operating systems installed. This will give the student a better chance of success to finding opens or vulnerabilities of hardware/computers with this course. Each victim station will be given a static IP announced to the student. This will save time as the potential hacker does spend a decent amount of time researching for IP addresses for their potential victims. A static IP will be given as there is potential of victim stations rebooting/etc. and or attaining different IP addresses which would only cause potential issues with students and their course progress if the IP addresses were dynamic.

Gateway: 192.168.90.1
192.168.90.255
Mask: 255.255.255.0
DHCP Range: 192.168.90.21-254

- **Victim 1: Windows PC**
IP: 192.168.90.18

USER: _____

PASS: _____

- **Victim 2: Windows Server**
IP: 192.168.90.19

USER: _____

PASS: _____

- **Victim 3: Linux**

IP: 192.168.90.20

USER: _____

PASS: _____

Why were these three specific operating systems (OS's) chosen for victim units? These are the least secure Operating Systems in 2014 as measured by the National Vulnerability Database (NVD) [1]. These were chosen so the students have a better chance for success at hacking the victim units.

Operating System	# of vulnerabilities	# of HIGH vulnerabilities	# of MEDIUM vulnerabilities	# of LOW vulnerabilities
Linux Kernel	119	24	74	21
MS Windows Server 2008	38	26	12	0
MS Windows 7	36	25	11	0
MS Windows	38	24	14	0

Serve r 2012				
MS Windo ws 8	36	24	12	0

Top Ten Least Secure Operating Systems

N e t w o r k

The three victims will be physical computers not virtual, however, a VLAN will be setup specifically for this victim network to allow students to access these victims from off campus.

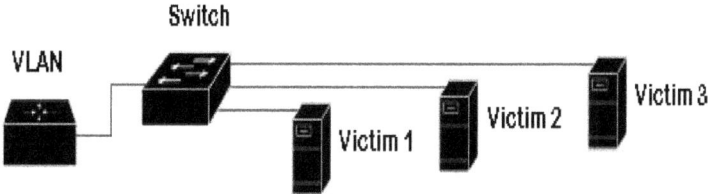

T u n n e l i n g

In past courses the three victims were physical computers. It is proposed that the new implementation of this course that the victims will still be virtual but be accessible to students both on and off campus through a network tunnel. The previous implementations off this course the student only had access to the victims if on campus and within a specific network lab/classroom. Additionally, through a physical hard wired connection between a student predator laptop and the three victims.

S o f t w a r e

No software in this course will be purchased. All software used by the student will be free. This freedom of thinking will give the student an edge in their journey of "thinking like a hacker". Hackers do not buy software. The thinking behind a successful hacker is to not be seen, traced or tracked and to hide behind hardware and or software and to give the illusion you are not who you are nor your intent. This can only be carried out through the attainment of free untraceable and non-serialized software.

T e x t b o o k s

The first textbook chosen for this course is titled "Penetration Testing Fundamentals". What better way to teach network personnel of offense network strategies than to use the structured thoughts and concepts used by Penetration Testers, often mentioned as Ethical Hackers? This textbook starts with the simple ideals and concepts behind some of the thinking of the potential hacker and goes into more depth as the chapter's progress.

ISBN: 978-0-7897-5937-5

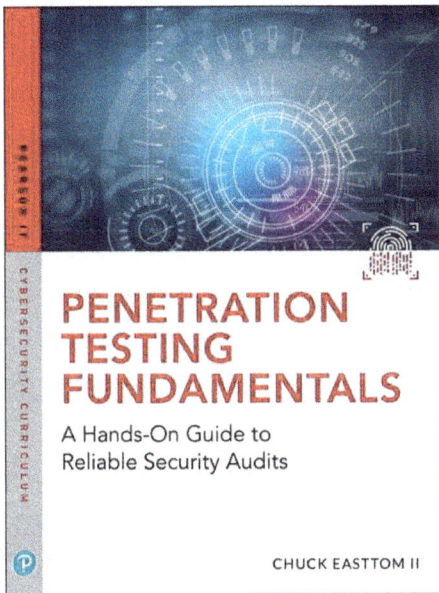

PENETRATION TESTING FUNDAMENTALS
A Hands-On Guide to Reliable Security Audits

CHUCK EASTTOM II

The second textbook chosen for this course is titled "Web Penetration Testing with Kali Linux". This textbook starts with the basics of the Linux operating system as well as some navigation and concepts behind some of the thinking of the potential hacker and goes into more depth within Linux as the chapter's progress.

ISBN: 978-1-78216-316-9

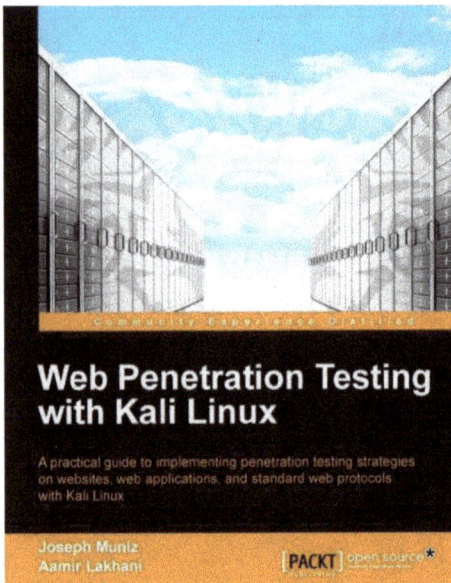

In addition to the two texts listed above you may find it necessary to search the internet for additional helpful materials for this course. Some of these are listed here:

- The Hacker Playbook 3: Practical Guide to Penetration Testing
- Common Ports.pdf
- Wireshark Display Filters.pdf
- 20 Fantastic Kali Linux Tools

- CEH Certified Ethical Hacker: Exam Guide
- Common Kali Linux Commands Overview
- Kali Linux Cookbook
- Penetration Testing Fundamentals: Coalfire
- Mastering Kali Linux for Advanced Penetration Testing
- Why Should I Care? Mobile Security for the Rest of Us
- Wireshark Quick Start Guide

Deliverables

The student will implement a structured portfolio of attack. The student will document the processes and procedures they use during this course daily, whether successful or unsuccessful. They will proceed through explanations of why they chose the software's they chose, what the software's told them, and if they thought the software's was successful for them. They will keep a daily attack diary of their successes and failures. The goal will be to successfully exploit a hole or vulnerability of one of the victim stations.

12 ... MY COURSE ACTIVITIES

In all of my courses I have activities I specifically make for student s to actually think and to find ways to try and relate something physically and or mentally to the student relative to the course topic.

Scenarios / Case Studies Activities

Let's look at a few penetration testing scenarios/case studies, each followed by ten questions. Can you answer the ten questions? See the wording of the questions how they differ for each type of test? They are similar but they have different terminologies for each type. Using the correct terminology is critical.

1 : Jacks Custom Targets (JCT)

Your team has been tasked with testing a corporate intranet (Jacks Custom Targets Inc.) that hosts sensitive employee information and internal communications tools. JCT is a company that makes custom 3D targets for government and military sniper schools. As I have mentioned many times in my courses, governments want to know that their potential vendors are as secure as they think they are or they will not do business with them. JCT is proposing to expand their relationship with the US government and wants to make sure their network environment is as secure as it can be.

Instructions

Answer the following questions with as much detail to get your point across to your team that you know how to do this task and they will have no legal or other ramifications when the task is complete.

Questions:

1. What tools will you use for network enumeration within the intranet?
2. How will you identify potential misconfigurations in internal servers?
3. What steps will you take to test for SQL injection vulnerabilities in internal applications?
4. How will you assess the effectiveness of existing firewall rules?
5. What are the implications of exploiting vulnerabilities in internal applications?
6. How will you validate the security of file-sharing services on the intranet?
7. What methods will you use to perform social engineering tests on employees?
8. How will you prioritize the vulnerabilities you discover?
9. What legal or ethical considerations will you address during your testing?
10. How will you present your findings to the IT department?

2: Franks Crypto Bank (FCB)

Your team has been contracted to test a mobile application for a banking institution (Franks Crypto Bank Inc.) that allows users to manage their accounts. FCB is on a list of crypto banking institutions the US government has developed to catch cyber criminals. Similar situations like this could be called a honey pot. A honey pot is where potential bees/flies (criminals) come to trade and are caught doing illegal activity they believe to be legitimate.

Instructions

Answer the following questions with as much detail to get your point across to your team that you know how to do this task and they will have no legal or other ramifications when the task is complete.

Questions:

1. What platforms (iOS/Android) will you focus on, and why?
2. How will you assess the app for insecure data storage practices?
3. What techniques will you use to analyze API calls made by the app?
4. How will you identify vulnerabilities related to user authentication?
5. What tools will you use to reverse-engineer the mobile application?
6. How will you test for code injection vulnerabilities in the app?

7. What are the risks of testing in a live environment?
8. How will you simulate man-in-the-middle attacks on the app?
9. What privacy concerns should you address regarding user data?
10. What steps will you take to document your findings?

3 : Y e e H a w   I n c .

A new rodeo management company (YeeHaw Inc.) has transferred its infrastructure to a cloud provider and wants an assessment of its configuration and services. They would like to know if they are secure as they have been hacked twice before and spend several million dollars on law suits in the past. They understand there is risk involved and they need to take additional steps to be more secure. Through their research they decided to spend money on the cloud instead of taking the risk, as they had in the past, with physical devices under their own protection, they never worked very well in the past. They are hoping this upgrade will be more secure.

I n s t r u c t i o n s

Answer the following questions with as much detail to get your point across to your team that you know how to do this task and they will have no legal or other ramifications when the task is complete.

Questions:

1. What cloud services and configurations will you review during your assessment?
2. How will you check for misconfigured security groups or firewalls?
3. What tools will you use to analyze IAM (Identity and Access Management) permissions?
4. How will you assess the security of data stored in cloud storage?
5. What steps will you take to test for vulnerabilities in server-less functions?
6. How will you evaluate third-party integrations and their security?
7. What are the risks associated with using default configurations in cloud environments?
8. How will you document any compliance issues you discover?
9. What remediation strategies will you recommend for vulnerabilities found?
10. How will you ensure the testing does not disrupt any ongoing services?

4 :   F U B A R   W i r e   I n c .

A premier wireless access company (FUBAR Wire Inc.) has requested a pen test of its wireless network to identify vulnerabilities that could be exploited by an attacker. FUBAR Wire is a subsidiary of a world-wide company that would like testing done to make sure their smaller entity is as safe as they feel they are.

I n s t r u c t i o n s

Answer the following questions with as much detail to get your point across to your team that you know how to do this task and they will have no legal or other ramifications when the task is complete.

Questions:

1. What tools will you use to scan for wireless networks and their configurations?
2. How will you assess the strength of the encryption used on the wireless network?
3. What methods will you employ to test for rogue access points?
4. How will you conduct a man-in-the-middle attack in a controlled manner?
5. What techniques will you use to capture and analyze wireless traffic?
6. How will you identify weak passwords used for the wireless network?
7. What steps will you take to ensure testing does not disrupt legitimate users?
8. How will you verify the security of guest networks versus internal networks?
9. What are the legal implications of testing a wireless network?
10. How will you report findings to ensure actionable insights for network administrators?

5 : M o o g l y

You are conducting a security assessment of a newly developed web application (Moogly) for a Veterans Administration healthcare provider, focusing on patient data management for PTSD military survivors and their families. They understand the necessity of security for the data they acquire. They are also an entity that works parallel with US governmental agencies so they are tasked with laws, regulations, and requirements that must be followed or penalties will ensue.

Instructions

Answer the following questions with as much detail to get your point across to your team that you know how to do this task and they will have no legal or other ramifications when the task is complete.

Questions:

1. What specific OWASP Top Ten vulnerabilities will you prioritize during your assessment?
2. How will you conduct a security review of the web application's source code?
3. What tools will you use for automated scanning of the web application?
4. How will you test for Cross-Site Scripting (XSS) vulnerabilities?
5. What techniques will you use to evaluate session management and authentication processes?
6. How will you check for proper implementation of HTTPS across the application?
7. What methods will you use to test for sensitive data exposure?
8. How will you validate the effectiveness of the web application's error handling?
9. What remediation suggestions will you provide for identified vulnerabilities?
10. How will you ensure your findings comply with industry regulations such as HIPAA?

Discussion Activities

In all of my courses I have discussion activities. These are made for students to give their ideas and or opinions on certain topics. Once they give their opinion other students will then join their discussion.

1. Regret

Have YOU ever known anyone that did something they should have not done? Maybe something that they may have regretted doing? All human entities have a past and most have a future. We all do stupid things in our lives. Some of us live to tell about them. Some of us could write a book about them as we have done so many, too many to list here. I can say for certain that I have and I should probably not be alive today. I always say I have nine lives.

Example: When I was about 2 years old. I went to the store with my mother. I saw this shiny object hanging on a hook. It was unusual to me. I put the item in my pocket. When I got home I pulled out the item and started playing with it. My mother asked me where I got it and I said at the store. She immediately took me back to the store and made me hand it to the clear, tell him what I had done, and apologize. Of course, corporal punishment was enlisted at my house, so I got a spanking I am sure. I would guess the spanking made me regret what I had done?

Have YOU ever done something YOU should have not done? If YOUR plan is to become involved in a security related career, make it a point, NOT to mess up those chances by doing something YOU may regret later that might keep YOU from YOUR goal! If YOU get a negative record with a law enforcement agency it alone may hinder YOUR intentions!

This course YOU are in this semester is related to cyber security. So, it would make sense for a student enrolled in a cyber security related program and taking a cyber security related course would start to think about things related to security and or cyber security. Right?

This is college, YOUR time going to k-12 classrooms is over. It is time to think like an adult and come to the realizations of an adult. Put YOUR cell phone down and pay attention!

THINK ABOUT IT!

Human entities do things every day they regret. Some human entities more than others. The thing about tit is: If YOU do not start it in the first place YOU won't have to worry about it! If YOU do not steal a first time then YOU will not have to worry about it a second time! If YOU do not cheat the first time then YOU will not be worried about being caught! Some time's we do things we justify and minimize in our mind. We know it is wrong and we try it anyway. Maybe we get away with it? Sometimes our minds come back to haunt us and just can't let some things go. The lesson is: If YOU don't try it in the beginning then YOU do not have to worry about it!

Example: I have quite a few regrets. One I have quite often. My dad died several years ago. We were never really close. He was a staunch disciplinarian. He was a military man. He was big and strong. I came into contact with his belt more times than I care to remember. I was not a good kid by any means. I guess I wanted more attention? My dad was always working. He never took me to a sports game. We never passed a ball around in the yard. Dad was always working and mom was the homemaker. Dad would go work in another state for months or years and return home with presents for us kids or we would follow him to the new state after a time. We moved 27 times before my senior year of high school. Anyway, I never realized when I was younger why he was the way he was, why he was never around, etc. I didn't realize until I got older! Some things I just did not realize until after his death, why he was the way he was. When WE both got older I never really went out of my way to befriend him and at least let my guard down and try. This still bothers me to this day. Maybe I could have reached out and done something? I blame myself for part of that!

Here is YOUR activity................

Think about what YOU have just read. There is no right or wrong answer here. No one on the planet is perfect. Everyone has a past and has done something we can look back on and say we should have not done that! YOU do not have to be specific, just think about how YOU want to answer this activity. If YOU have no specific issue to spell out then maybe someone YOU know has?

The activity is to write up and submit an essay for credit points. YOU will use a 10 point font, correct spelling, grammar, full sentences and structure.

Your initial reply to the prompt below should be at minimum <u>200+ words, ideally three paragraphs</u>. Be sure that your post provides a clear and concise response to the prompt, provides support when needed, and engages your classmates. You will not be able to view other students' posts prior to submitting your own post.

You must reply to at least <u>two (2) classmates'</u> posts with <u>more than 100 words per reply</u>. These replies must *further the discussion* by providing an alternative point or asking a further question while providing your own perspective. Always be sensitive to the other person's perspective, even if it differs from your own.

Initial posts are due by 11:59 pm on Wednesday of this week. Replies are due by 11:59 pm on Sunday of this week.

2 . Struggle

There is a saying that has been around for many years. I am not sure where it came from but I have heard it all my life. "What doesn't kill YOU makes YOU stronger."

In life, it is a fact that, human entities are born with no knowledge, a blank slate. Our brains are empty. We have never been able to use our senses to gain knowledge which then would transfer to our brains and therefore, learn. YOU cannot learn from experiences unless YOU have them! YOU cannot gain positive experiences unless YOU come into contact with them, and vice-versa for negative experiences.

Another saying: "We are products of our environments." From the day we are born, human entities come out into the world seeking interaction. If YOU have ever been around a little child, it seems the only word they know is "Why"? Why, Why, Why, Why, and Why? It seems it is a never-ending question.

Every situation we come in contact with during our lives on this planet we gain knowledge. We really don't realize most of the time, but our brain is working 24/7 to process information our senses give it. This happens from when we awake for our day, until we go to bed, and during our naps and sleep. Some of our thoughts and processes are slower and some of ours are faster. Whether YOU realize it or not YOU are a product of YOUR environment. YOU are a product of everything YOU have come into contact with since the day YOU were born.

This course YOU are in this semester is related to cyber security. So, it would make sense for a student enrolled in a cyber security related program and taking a cyber security related course would start to think about things related to security and or cyber security. Right?

This is college, YOUR time going to k-12 classrooms is over. It is time to think like an adult and come to the realizations of an adult. Put YOUR cell phone down and pay attention!

THINK ABOUT IT!

Example: YOU are a product of everything YOU have come into contact with since the day YOU were born. We all have positive and negative things that happen in our lives. Positives generally build us up and negatives tend to tear us down. Positives give us motivation and negatives give us question. Another saying I have heard my whole life: "Bad things happen to good people and good things happen to bad people." It is relatively a never ending cycle human entities go through each day. Sometimes new things happen and sometimes they repeat. Some good and some bad. There is no such thing as perfection. If there was, we would have good happen all the time. This isn't realistic. YOU have to accept the good with the bad.

Here is YOUR activity...............

Motivation can come in many forms...Struggles happen to all of us. PICK ONE thing, either positive or negative, that has happened during YOUR life that has motivated YOU to seek an education. In addition, add answers to these questions:

What motivated you?
How will that motivation keep YOU on track?

The purpose of this assignment is to be aware of YOUR motivation! To keep YOUR motivation! To not let anything remove YOUR motivation! Motivation is success! YOU want to achieve success, right?

The activity is to write up and submit an essay for credit points. YOU will use a 10 point font, correct spelling, grammar, full sentences and structure.

I n s t r u c t i o n s

Your initial reply to the prompt below should be at minimum 200+ words, ideally three paragraphs. Be sure that your post provides a clear and concise response to the prompt, provides support when needed, and engages your classmates. You will not be able to view other students' posts prior to submitting your own post.

You must reply to at least two (2) classmates' posts with more than 100 words per reply. These replies must *further the discussion* by providing an alternative point or asking a further question while providing your own perspective. Always be sensitive to the other person's perspective, even if it differs from your own.

Initial posts are due by 11:59 pm on Wednesday of this week. Replies are due by 11:59 pm on Sunday of this week.

O r a l / P o w e r P o i n t A c t i v i t i e s

In all of my courses I have activities where students create a PowerPoint slide show. Students may be asked to give these as oral presentations to the class.

1 : T e a m A c t i v i t y " I n t h e N e w s "

We see the news, we read the news, and we hear the news. We get the news on our phones, on our televisions, and on our computers. Sometimes we don't quite realize the relevance on news stories in some of the courses we take throughout our educational careers.

In this Activity/Assignment you will think about topics you read about in this courses and how those topics may relate to things we hear and see in the news. You will point out 1 article in the news that is relative to topics in this course. You will put together a PowerPoint presentation to share with the class. It will be submitted for credit and may be presented to the class.

Instructions

The presentation will consist of a title page, 4 to 5 pages of content, and a final reference page (Where you got the story/article). The pages can be as complex as you like or as simple as you like.

For this credit, you MUST attach your completed PowerPoint as a MS Office Suite PowerPoint or you can export as an Adobe PDF.

Title page (Your name, course, semester, "In the News")

Reference page (You will have a reference page from where you got the story/article)

Check ALL spelling and grammar before submission.

***DO NOT use a Power Point previously submitted in another course! It will be zero credit if so.

For a total of 100% for accurately completing this activity by the due date.

2: Solo Activity "Five Things I Learned"

Some students don't quite realize that they actually have learned something/somethings in their classes. I don't understand how a student can say this when they have passed a course, let alone pass the course with a good grade?

In this Activity/Assignment you will think about what you have learned in this course. You will point out 5 items you can truly say you have learned. You will put together a PowerPoint presentation to share with the class. It will be submitted for credit and may be presented to the class.

Instructions

The presentation will consist of a title page, 4 to 5 pages of content, and a final reference page if needed. The pages can be as complex as you like or as simple as you like.

For this credit, you MUST attach your completed PowerPoint as a MS Office Suite PowerPoint or you can export as an Adobe PDF.

Title page (Your name, course, semester, "Five Things I have Learned in this Course")

Reference page (You will have a reference page if necessary)

Check ALL spelling and grammar before submission.

***DO NOT use a Power Point previously submitted in another course! It will be zero credit if so.

For a total of 100% for accurately completing this activity by the due date.

Thinking Assignment Activities

In all of my courses I have what I call thinking activity assignments. I generally give students a scenario and ask them how they might resolve that scenario or give their opinion as to how to resolve.

Instructions

Explain with details using a minimum of 200+ words, about 3 paragraphs. Use at least APA version 7 to provide evidence from and cite your textbook and at least one high-quality internet source. Be sure that your response is clear and concise and provides adequate support when needed.

Answer the following:

....
....
....

- Attach your completed paper as a .doc, .docx, or .pdf file.
- Check all spelling and grammar before submission.

*DO NOT use a paper previously submitted in another course. This is plagiarism and will result in an "F" in the course.

This assignment is due by 11:59 pm on Sunday of this week.

Essay Assignment Activities

In all of my courses I have written essay assignments. These entail a little more research than the previous assignments. Here again I try to make the students think and support their thoughts with references from the field.

Instructions

Write a 250-500 word essay, focusing on the certification you chose in the Certifications and Marketing assignment. Discuss whether employers agree with the marketing and advertising and find it useful and viable. This perspective is valuable because employers are the consumers of these certifications rather than the marketers of them. Use APA 7 to provide evidence from and cite any sources you may use.

- Attach your completed paper as a .doc, .docx, or .pdf file.
- Check all spelling and grammar before submission.

*DO NOT use a paper previously submitted in another course. This is plagiarism and will result in an "F" in the course.

This assignment is due by 11:59 pm on Sunday of this week.

1 3 ... O U R   N E T W O R K   L A B

In this course we will be using a laptop physically connected by a cable to a switch physically connected to three servers: One MS Windows, One MS Windows Server, and One Linux.

Within our Network Lab:

* Color-Coded Ports for Physical Connectivity
* Four Rows of Five PCs (Connected to Physical Campus Network: Not to be used as Predators)
* Each Row Connects to One of Four Racks (Alpha, Bravo, Charlie, and Delta)

In the image below you will see the typical 'Wired network.

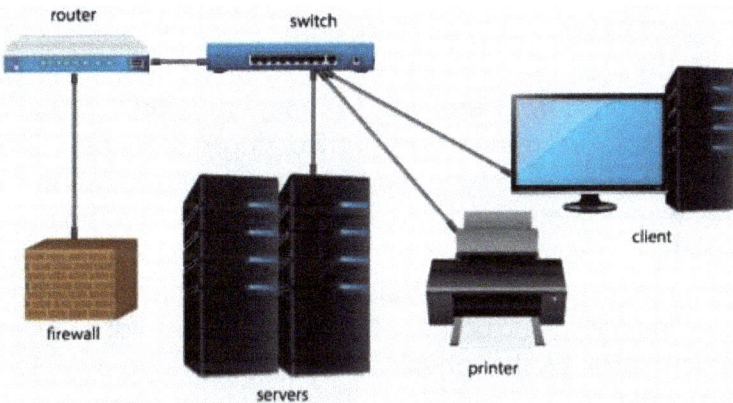

Generic Image: Typical Wired Network

In the image below you will see the typical 'Wireless network. Although, with newer technologies becoming available all the time, you may see some of these change or become more advanced.

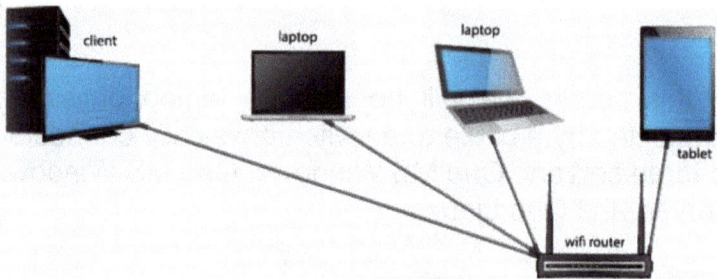
Generic Image: Typical Wireless Network

Network Topologies

A network topology is the basic design of a computer network, much like a road map. This map details how key network components such as nodes and links are interconnected. It is comparable to the blueprints of a house in which components such as electrical systems, HVAC systems, and plumbing systems are integrated into the overall design.

Taken from the Greek word 'Topos' meaning "Place", Topology, in relation to networking, describes the configuration of the network, including the location of the workstations and wiring connections. A topology, which is a pattern of interconnections among nodes, influences a network's cost and performance. There are basically four primary types of network topologies which refer to the physical and logical layout of the network cabling. The Mesh, Star, Bus and Ring.

"Knowing the network topology of your victim or victims can be most useful in breaching their network."

Of the four types of network topologies, the "Mesh" is the most prevalent one and the one you should understand.

M e s h

In a mesh topology, every device has a dedicated point-to-point link to every other device. The term *dedicated* means that the link carries traffic only between the two devices it connects.

S t a r

All devices connected with a Star setup communicate through a central Hub by cable segments. Signals are transmitted and received through the Hub. It is the simplest and the oldest and all the telephone switches are based on this. In a star topology, each network device has a home run of cabling back to a network hub, giving each device a separate connection to the network. So, there can be multiple connections in parallel.

B u s

The simplest and one of the most common of all topologies, a Bus consists of a single cable, called a Backbone that connects all workstations on the network using a single line. All transmissions must pass through each of the connected devices to complete the desired request. Each workstation has its own individual signal that identifies it and allows for the requested data to be returned to the correct originator.

In the Bus Network, messages are sent in both directions from a single point and are read by the node (computer or peripheral on the network) identified by the code with the message. Most Local Area Networks (LANs) are Bus Networks because the network will continue to function even if one computer is down.
This topology works equally well for either peer to peer or client server.

Ring

All the nodes in a Ring Network are connected in a closed circle of cable. Messages that are transmitted travel around the ring until they reach the computer that they are addressed to, the signal being refreshed by each node. In a ring topology, the network signal is passed through each network card of each device and passed on to the next device. Each device processes and retransmits the signal, so it is capable of supporting many devices in a somewhat slow but very orderly fashion. There is a very nice feature that everybody gets a chance to send a packet and it is guaranteed that every node gets to send a packet in a finite amount of time.

Mesh - Star - Bus - Ring ?

Generally, a BUS architecture is preferred over the other topologies – of course, this is a very subjective opinion and the final design depends on the requirements of the network more than anything else. Lately, most networks are shifting towards the STAR topology. Ideally we would like to design networks, which physically resemble the STAR topology, but behave like BUS or RING topology.

What about Cables

There are quite a few cables in our Network Lab. What are they for? Are they all the same?

All cables are not the same. All cables do not have the same plugs on the ends or the same way to connect them to other devices. The makeup or composition of the cables are different.

Let's start by saying there are three basic connections to devices: One is a physical copper wired connection, which can be with either solid copper wires or stranded. One is a fiber-optic glass connection, which can be either single-mode or multi-mode. The last connection can be a wireless connection which in actuality has no physical connectivity at all!

Although, I will say this: EVERY connection has some form of physical connection at some point in the network! When you talk of wireless connectivity, there is NO such thing as a 100% wireless network. It does not exist. Somewhere in that 'chain' is some form of a 'wired' connection.

When talking about cable it would be incomplete if we did not mention cross-over cables. Cross-over cable are used to attach one single computer to another single computer. However, technologies are changing from some of the ports we used in the past. To explain a crossover is best to use an image. A picture is worth 1000 words, right?

Below is an image of a 'Phone' plug end and an 'Ethernet' plug end. Notice the 'Phone' plug has less wires than the 'Ethernet' plug. The 'Ethernet' plug has 8 wires. This is the only one we are interested in during this course.

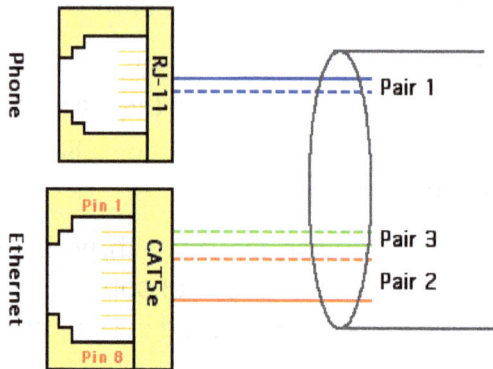

CAT-3 wire for phone plus 10base-t/T568A

Generic Image: CAT-3 and CAT5/6 Plug Wire Color Arrangement

How about another image of an 'Ethernet' plug.

Now let's get back to the cross-over cable we were talking about. In a cable there can be multiple wires, sometimes they are slid wires and sometimes they are stranded, depending upon their application. Without getting too complicated and confusing: Within the network cabling environment I will say solid cables are for the infrastructure and the stranded are for patching. If that makes sense? It would be difficult to punch down a patch panel with stranded cables, therefore they are solid. It could be problematic to use solid cables as patch cables as they are always being moved and twisted around.

There are two wiring standards for 'Ethernet' cables. This has to do with how the wires attach on each end of the cable or plug into the plug on each end. If that makes sense? The two standards are: TIA/EIA 568A & TIA/EIA 568B. In the image below you will see both wiring standards. On the left side you will see those as well as on the right side you will see the crossed wiring versions, hence the term cross-over cables.

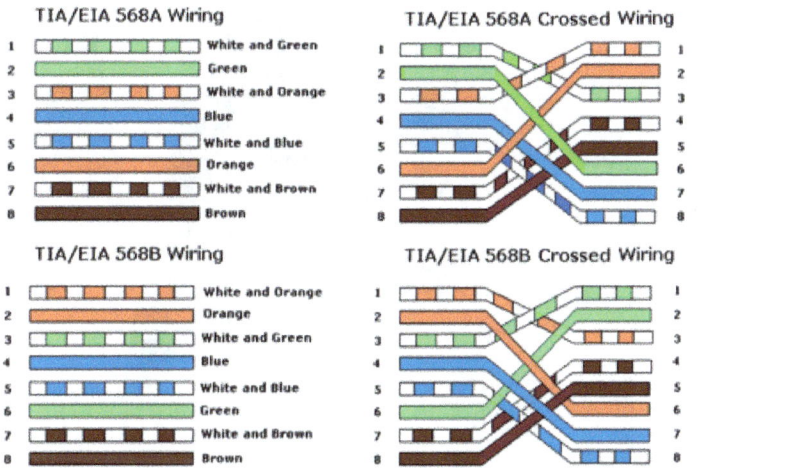

TIA/EIA 568A Wiring		TIA/EIA 568A Crossed Wiring
1 White and Green		1 ... 1
2 Green		2 ... 2
3 White and Orange		3 ... 3
4 Blue		4 ... 4
5 White and Blue		5 ... 5
6 Orange		6 ... 6
7 White and Brown		7 ... 7
8 Brown		8 ... 8

TIA/EIA 568B Wiring		TIA/EIA 568B Crossed Wiring
1 White and Orange		1 ... 1
2 Orange		2 ... 2
3 White and Green		3 ... 3
4 Blue		4 ... 4
5 White and Blue		5 ... 5
6 Green		6 ... 6
7 White and Brown		7 ... 7
8 Brown		8 ... 8

Figure A   Figure B

Shows the Pin Out of Straight through Cables   Shows the Pin Out of Crossover Cables

Generic Image: CAT5/6 568A/B Plug Wire Color Arrangement

If you notice all pinouts are called pairs. These have to do with the coloring of the wires. You will also notice in this pinout ONLY pairs #2 and #3 are actually used.

The other wires are punched down and connected, but nothing travels through these other two pairs.

T y p i c a l   N e t w o r k   T o o l s

Just to give a little more detail when I talk about cables and cabling below is an image of the typical equipment used to make and test network related cabling.

Generic Image: Typical Useful Tools of the IT Networking Trade

F i b e r - O p t i c s   i n   B r i e f

Just to mention and show an image. Below is an image of the inside of a typical Fiber-Optic cable.

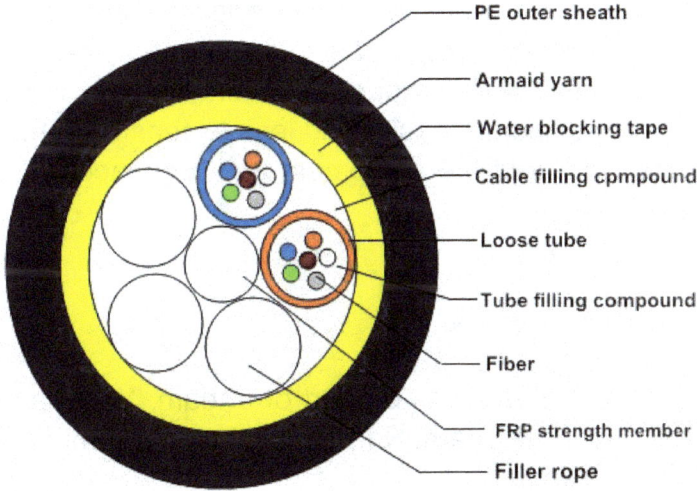

Generic Image: Typical Layout of a Fiber Optic Cable

Also, to add: I mentioned earlier the two types of Fiber-Optic cables Single-mode and Multi-mode.

Below is an image of the differences between modes.

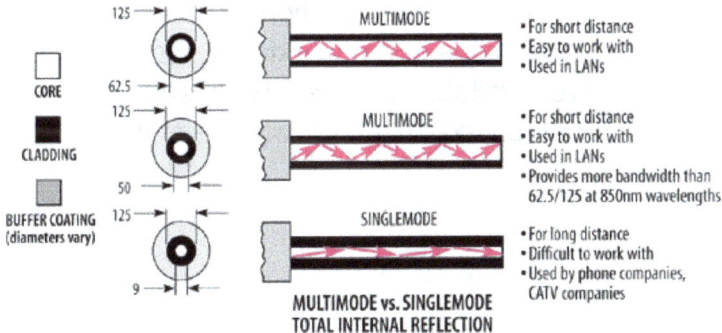

Generic Image: Fiber Optic Mode Differences

You probably have a few questions related to Fiber-optic and the difference of using it and copper?

To give a brief answer: Fiber-optic cables transfer longer distances with less loss of copper. Fiber-optic cables are more expensive then copper cables. Copper cables are easier and less expensive to repair in the field, and the equipment and tools needed are cheaper.

T e a m w o r k

You may be asked to be part of a team. You may be assigned: Blue Team, Red Team, Green Team, and or Yellow Team.

It is YOUR responsibility to be a viable part of the team! Team members will be asked about each of the participants within that team for grading purposes.

*If a team is assigned, each team will submit a team portfolio for submission each seek. Each participant will keep their own portfolio to later be used after graduation if needed. Penetration Testers are highly sought after positions in the workforce. Some of those positions are available at over 100k salary with no degree other than extensive knowledge of penetrating.

14 ... PEN TESTING: COMMAND-LINE TOOLS

Now that I have YOUR attention let's look at a significant part of Cyber Security: Offense. Cyber Security is not just about defense, it is about offense as well, and what better way to understand Cyber Security than through understanding offensive and defensive techniques through what we call: Penetration Testing.

Penetration testing involves using a variety of tools and techniques to identify and exploit vulnerabilities in systems, networks, and applications. Command-line tools are often preferred by penetration testers due to their flexibility, efficiency, and scripting capabilities. Here are some common command-line tools used in penetration testing:

N m a p : A powerful network scanner used for host discovery, port scanning, service version detection, and OS fingerprinting.

M e t a s p l o i t   F r a m e w o r k : An advanced exploitation framework that provides a wide range of exploit modules, payloads, and auxiliary modules for penetration testing.

S q l m a p : An automated SQL injection tool for detecting and exploiting SQL injection vulnerabilities in web applications.

H y d r a : A fast and flexible password-cracking tool that supports various protocols, including SSH, FTP, HTTP, and more.

John the Ripper: A popular password-cracking tool known for its ability to crack password hashes using various attack modes and techniques.

Burp Suite: A comprehensive web application security testing tool that includes a proxy, scanner, intruder, repeater, sequencer, and decoder.

Wireshark: A network protocol analyzer that allows you to capture and inspect the data flowing over a network in real-time.

Aircrack-ng: A set of tools for auditing wireless networks, including packet capture, WEP and WPA/WPA2-PSK cracking, and more.

Dirb or Dirbuster: Web directory brute-forcing tools used to discover hidden directories and files on web servers.

Gobuster: A directory and file brute-forcing tool similar to Dirb and Dirbuster but with faster performance and additional features.

Netcat (nc): A versatile networking utility that can be used for port scanning, banner grabbing, creating reverse shells, and transferring files.

Tcpdump: A command-line packet analyzer that allows you to capture and analyze network traffic.

Snort: An open-source intrusion detection system (IDS) that can be used for real-time traffic analysis and packet logging.

Hashcat : An advanced password recovery tool that supports various hash algorithms and attack modes for cracking password hashes.

Nikto : A web server scanner that checks for known vulnerabilities, misconfigurations, and outdated software in web servers.

Sqlninja : An exploitation tool for SQL injection vulnerabilities, providing post-exploitation functionality such as file retrieval and command execution.

Recon-ng : A full-featured reconnaissance framework that provides a powerful command-line interface for conducting reconnaissance and information gathering.

BloodHound : A tool for analyzing Active Directory environments to identify attack paths and potential security vulnerabilities.

Empire : A post-exploitation framework that provides a PowerShell-based agent for conducting various offensive operations on compromised systems.

CrackMapExec (CME) : A post-exploitation tool for Windows systems that allows you to execute commands, dump credentials, and perform lateral movement.

These are just a few examples of the many command-line tools available for penetration testing. Depending on the specific objectives and requirements of a penetration test, penetration testers may use a combination of these tools to identify, exploit, and mitigate security vulnerabilities.

In penetration testing, particularly during the reconnaissance and exploitation phases, you might need to utilize various command-line tools to gather information, perform network analysis, and exploit vulnerabilities. Here are some common Windows and Linux commands used for penetration testing:

15 ... PEN TESTING: WINDOWS AND LINUX COMMANDS

i p c o n f i g : Displays the TCP/IP network configuration including: IP address, subnet mask, MAC address, and default gateway.

n e t s t a t : Displays active network connections, listening ports, interface statistics, masquerade connections, multicast memberships, and routing tables. Useful for identifying network activity and potential vulnerabilities.

n s l o o k u p : Command-line tool for querying DNS servers to obtain domain name or IP address information.

t a s k l i s t : Lists all running processes on a Windows system. Useful for identifying running services and potential targets for exploitation.

n e t : Command-line tool for managing network resources, including users, groups, shares, and network connections.

r o u t e : Displays and modifies the IP routing table. Useful for viewing and managing network routes.

t r a c e r t : Traces the route taken by packets across an IP network. Useful for identifying network hops and diagnosing network connectivity issues.

a r p : Displays and modifies the Address Resolution Protocol (ARP) cache, which maps IP addresses to MAC addresses.

p i n g : Sends ICMP echo requests to a target host to test network connectivity and measure round-trip latency.

t a s k k i l l : Terminates one or more running processes on a Windows system.

L i n u x   O n l y

p s : Lists information about currently running processes.

k i l l : Sends a signal to a process, allowing you to terminate it.

d i g : A versatile DNS lookup utility for querying DNS name servers.

16 ... PEN TESTING: MOST USED PROMPTS

These are just a few examples of commonly used commands in penetration testing. Depending on the specific task or objective, penetration testers may use a wide range of command-line tools to gather information, analyze networks, and exploit vulnerabilities.

In MS Windows go to the bottom of the desktop to the 'Taskbar' at the bottom. There is usually a space there that has a little magnifying glass symbol and says "Type here to search". In that space type "CMD" then hit enter. This will bring up the BLACK Command Prompt box.

Below is a screen shot of what you might see?

```
Command Prompt                        —    □    ×
Microsoft Windows [Version 10.0.19045.3930]
(c) Microsoft Corporation. All rights reserved.

C:\Users\Cwiseguy>_
```

ipconfig

In computing, 'ipconfig' in MS Windows is a console application that displays all current TCP/IP network configuration values and can modify Dynamic Host Configuration Protocol and Domain Name System settings. Sometimes you have to 'release' or 'renew' an IP address for your system as it may conflict with another on the network with the same IP address. Why you ask? Well, there are static IP addresses and dynamic IP addresses. A static IP is when you give a network device and IP that will stay the same and never change. A dynamic IP address is one that changes or may change each time it connects to a network. There are devices on a network that assign IP addresses, as in dynamic IP addresses.

Below is a screenshot of what you might see?

p i n g

PING is a computer network administration software utility used to test the reachability of a host on an Internet Protocol (IP) network.

Below is a screenshot of what you might see?

```
C:\>ping 192.168.1.224

Pinging 192.168.1.224 with 32 bytes of data:
Reply from 192.168.1.224: bytes=32 time<1ms TTL=128
Reply from 192.168.1.224: bytes=32 time<1ms TTL=128
Reply from 192.168.1.224: bytes=32 time<1ms TTL=128
Reply from 192.168.1.224: bytes=32 time<1ms TTL=128

Ping statistics for 192.168.1.224:
    Packets: Sent = 4, Received = 4, Lost = 0 (0% loss),
Approximate round trip times in milli-seconds:
    Minimum = 0ms, Maximum = 0ms, Average = 0ms

C:\>
```

t r a c e r t

Traceroute is a command which can show you the path a packet of information take's from your computer to another computer that you specify. It will list all the routers it passes through until it reaches its destination, or fails to reach and is discarded. In addition to this, it will tell you how long each 'hop' from router to router takes.

In MS Windows, you need to start out with a Command Prompt. "Which you will see those directions here in this writing also as a Tip."

Once in the window of a command prompt, ENTER the word 'Tracert' followed by a space, then the domain name you are reviewing.

Below is a screen shot of what you might see?

```
Command Prompt                                                         _|□|x|
C:\>tracert mediacollege.com

Tracing route to mediacollege.com [66.246.3.197]
over a maximum of 30 hops:

  1    <10 ms    <10 ms    <10 ms  192.168.1.1
  2    240 ms    421 ms     70 ms  219-88-164-1.jetstream.xtra.co.nz [219.88.164.1]
  3     20 ms     30 ms     30 ms  210.55.205.123
  4       *         *         *    Request timed out.
  5     30 ms     30 ms     40 ms  202.50.245.197
  6     30 ms     40 ms     40 ms  g2-0-3.tkbr3.global-gateway.net.nz [202.37.245.140]
  7     30 ms     30 ms     40 ms  so-1-2-1-0.akbr3.global-gateway.net.nz [202.50.116.161]
  8    160 ms    161 ms    160 ms  p1-3.sjbr1.global-gateway.net.nz [202.50.116.178]
  9    160 ms    171 ms    160 ms  so-1-3-0-0.pabr3.global-gateway.net.nz [202.37.245.230]
 10    160 ms    161 ms    170 ms  pao1-br1-g2-1-101.gnaps.net [198.32.176.165]
 11    180 ms    181 ms    180 ms  lax1-br1-p2-1.gnaps.net [199.232.44.5]
 12    170 ms    170 ms    171 ms  lax1-br1-ge-0-1-0.gnaps.net [199.232.44.50]
 13    240 ms    241 ms    240 ms  nyc-n20-ge2-2-0.gnaps.net [199.232.44.21]
 14    240 ms    251 ms    250 ms  ash-n20-ge1-0-0.gnaps.net [199.232.131.36]
 15    241 ms    240 ms    250 ms  0503.ge-0-0-0.gbr1.ash.nac.net [207.99.39.157]
 16    251 ms    260 ms    250 ms  0.so-2-2-0.gbr2.nyr.nac.net [209.123.11.29]
 17    250 ms    260 ms    261 ms  0.so-0-3-0.gbr1.oct.nac.net [209.123.11.233]
 18    250 ms    260 ms    261 ms  209.123.182.243
 19    250 ms    260 ms    261 ms  sol.yourhost.co.nz [66.246.3.197]

Trace complete.

C:\>
```

In this example: Firstly it tells you that it's tracing the route to mediacollege.com, tells you the IP address of that domain, and what the maximum number of 'hops' will be before it times out.

Next it gives information about each router it passes through on the way to its destination:

Line 1 is the internet gateway on the network this traceroute was done from (an ADSL modem in this case)
Line 2 is the ISP the origin computer is connected to (xtra.co.nz)
Line 3 is also in the xtra network
Line 4 timed out
Lines 5 - 9 are all routers on the global-gateway.net.nz network (the domain that is the

internet gateway out of New Zealand)
Lines 10 - 14 are all gnaps.net in the USA (a telecom supplier in the USA)
Lines 15 - 17 are on the nac network (Net Access Corporation, an ISP in the New York area)
Line 18 is a router on the network mediacollege.com is hosted on and finally
Line 19 is the computer mediacollege.com is hosted on (sol.yourhost.co.nz)

W H O I S

WHOIS (pronounced as the phrase *who is*) is a query and response protocol that is widely used for querying databases that store the registered users or assignees of an internet resource, such as a domain name, an IP address block, or an autonomous system, but is also used for a wider range of other information. The protocol stores and delivers database content in a human-readable format.

P u t t y

PuTTY is an open source software used as a SSH and or telnet client. Basically it is a terminal emulator. A free serial console network file transfer application. It is a free download and has been out for 23 years. There are some settings you need to figure out when using this program depending upon what you are using it for.

Be Careful

When you do something wrong you can be prosecuted. Sometimes a simple apology does not suffice. In the previous chapter you are given instructions. Follow those instructions to the letter. If those instructions mention paying attention to what network you are trying to break into, try to break into that network only! This is NOT one of those times to try and see how smart you are, and sneak into another network or the campus network, to see if you are being tracked. Trust me, when I say you are being tracked and forgiveness may not be an option. Public educational facilities are federal institutions which carry larger penalties than state facilities.

You will be completing and submitting a White-Hat Agreement form. It is to be completed and submitted the first week of this course.

17 ... PEN TESTING: COMMANDS FOR PHASES

Penetration testing, often abbreviated as "pen testing," involves simulating real-world cyberattacks to identify vulnerabilities in systems, networks, and applications. While there are numerous tools and techniques used in penetration testing, here are some common and useful commands for various phases of the process:

Phase 1 - Gathering Information:

n m a p :   Network mapping tool for discovering hosts and services on a network.

w h o i s :   Query WHOIS databases to gather information about domain registrations.

d n s e n u m :   DNS enumeration tool for gathering DNS information about a domain.

d i r b   o r   d i r b u s t e r : Web directory brute-forcing tools for finding hidden directories and files on web servers.

t h e H a r v e s t e r :   Email, subdomain, and employee name harvesting tool.

Phase 2 - Assessing Vulnerabilities:

nikto: Web server scanner that checks for known vulnerabilities, misconfigurations, and outdated software.

OpenVAS or Nessus: Vulnerability scanners for identifying vulnerabilities in networks and web applications.

nmap --script vuln: Nmap script for detecting vulnerabilities on target hosts.

Metasploit Framework: Exploitation framework with a vast database of exploits for testing vulnerabilities.

Phase 3 - Exploitation:

searchsploit: Searchable database of exploits and vulnerabilities to find exploits for known vulnerabilities.

exploitdb: Exploit Database containing exploits, shellcodes, and papers for penetration testers.

msfconsole: Command-line interface for the Metasploit Framework for exploiting vulnerabilities.

sqlmap: Automatic SQL injection and database takeover tool.

Phase 4 - Post-exploitation:

netcat (nc): Swiss army knife for network connections, useful for creating reverse shells or transferring files.

Meterpreter (part of Metasploit): Advanced, multi-function payload that can be injected into exploited systems for post-exploitation tasks.

enum4linux: Tool for enumerating information from Windows and Samba systems, useful for post-exploitation on Windows networks.

Phase 5 - Privilege Escalation:

LinEnum or Linux Privilege Escalation Checker: Scripts for checking common privilege escalation vectors on Linux systems.

windows-privesc-check or PowerUp: Scripts for checking common privilege escalation vectors on Windows systems.

sudo -l: Check sudo privileges for the current user.

g e t s y s t e m (M e t e r p r e t e r c o m m a n d) : Attempt to escalate privileges on a compromised Windows system.

P h a s e 6 - C o v e r i n g Y o u r T r a c k s :

s h r e d : Securely delete files from a system.

h i s t o r y - c : Clear command history.

n e t c a t (n c) o r M e t a s p l o i t : Use reverse shells to maintain access without leaving logs on the compromised system.

18 ... FREE PROGRAMS FOR HACKING AND PEN TESTING

**These are current AI recommendations as of this writing. Please use at your own risk. ONLY run and install these on the Predator laptops. You will have to search for these as they may have moved.*

Aircrack-ng: A suite of tools for assessing Wi-Fi network security. It includes tools for monitoring, attacking, and cracking WEP and WPA/WPA2 encryption keys.

Burp Suite Community Edition: A tool for web application security testing with a free version available.

Hack The Box: An online platform that provides a variety of cybersecurity challenges in a controlled environment.

Hydra: A fast and flexible tool for cracking login passwords using a dictionary-based approach. It supports many different protocols.

John the Ripper: A password cracking tool that supports various encryption algorithms and is useful for testing the strength of password hashes.

Kali Linux: A distribution designed for security professionals that includes many tools for testing and securing systems.

Metasploit Framework: An open-source tool used for developing and executing exploit code against a remote target machine. It's widely used by security professionals.

Netcat: Often referred to as the "Swiss Army knife" of networking, Netcat is used for a variety of network-related tasks, including port scanning and transferring files.

Nikto: A web server scanner that detects vulnerabilities and security issues in web servers by running a series of tests.

Nmap: A network scanning tool that can help identify devices on a network, discover open ports, and more.

OpenVAS: A vulnerability scanner that helps identify and manage security vulnerabilities in networked systems. It is part of the Greenbone Vulnerability Management (GVM) suite.

OWASP ZAP: An open-source tool designed for finding vulnerabilities in web applications.

SQLmap: An open-source tool for automating the detection and exploitation of SQL injection vulnerabilities in web applications.

TryHackMe: Another platform offering guided, hands-on cybersecurity training.

Wireshark: A network protocol analyzer that helps in capturing and analyzing network traffic.

**These tools and platforms are intended for ethical hacking and improving security practices. Penetration testing is a critical part of assessing and securing computer systems and networks. There are several free and open-source tools available that can help with different aspects of penetration testing.

19 ... HACKING BREACH LIBRARY

Hacking refers to unauthorized access, manipulation, or exploitation of computer systems, networks, or devices. Hacking techniques can vary widely in terms of their goals, methods, and the level of sophistication involved.

Ethical

Ethical hackers are authorized individuals or professionals who legally attempt to identify and fix vulnerabilities in systems, networks, or applications. They help organizations improve their security posture by finding and fixing security weaknesses before malicious hackers can exploit them.

Malware-Based

Viruses are malicious programs that infect computers and replicate themselves to spread to other systems. They can damage files, steal data, or give attackers control over infected computers. **Worms** are self-replicating malware that spreads across networks without human intervention, exploiting vulnerabilities in network protocols. **Trojans** are malware disguised as legitimate software. They appear harmless but perform malicious activities once installed, such as stealing sensitive information or providing remote access to attackers.

Phishing and Social Engineering

Phishing is sending deceptive emails, messages, or websites that trick recipients into revealing sensitive information like passwords or credit card numbers. **Spear Phishing** are targeted phishing attacks against specific individuals or organizations, using personalized information to increase credibility. **Baiting** is offering something enticing (e.g., free software download) to entrap users into clicking malicious links or downloading malware.

Denial-of-Service (DoS)

DoS Attack is overloading a system or network with excessive requests to render it unusable for legitimate users.

Distributed Denial-of-Service (DDoS)

DDoS Attack is coordinating multiple systems (botnets) to flood a target system or network with traffic, causing it to crash or become inaccessible.

Man-in-the-Middle (MitM)

Intercepting and eavesdropping on communication between two parties without their knowledge. Attackers can manipulate or alter messages, steal sensitive information, or impersonate legitimate users.

SQL Injection (SQLi)

Exploiting vulnerabilities in web applications by injecting SQL code into input fields. This allows attackers to manipulate databases, retrieve sensitive data, or execute unauthorized commands.

Cross-Site Scripting (XSS)

Injecting malicious scripts into web pages viewed by other users. When victims visit the compromised pages, the scripts execute in their browsers, allowing attackers to steal session cookies, redirect users to phishing sites, or deface websites.

Password

Brute Force Attack is trying all possible combinations of passwords until the correct one is found. **Dictionary Attack** is using a precompiled list of common passwords or phrases to guess a user's password. **Credential Stuffing** is Using previously leaked credentials from one service to gain unauthorized access to other accounts where users have reused passwords.

Internet of Things (IoT)

Targeting vulnerabilities in internet-connected devices such as smart cameras, routers, or thermostats to gain unauthorized access, launch attacks, or steal data.

Physical Security

Gaining unauthorized access to physical premises or systems by exploiting weaknesses in physical security measures, such as bypassing locks or tampering with hardware.

Advanced Persistence Threats (APTs)

Long-term targeted attacks by skilled and persistent adversaries who aim to steal sensitive information, disrupt operations, or maintain unauthorized access over an extended period without detection.

Crypto-Jacking

Illegally using other people's computing resources to mine cryptocurrencies without their consent or knowledge, often by infecting computers or websites with malware.

Wireless Hacking

Exploiting vulnerabilities in wireless networks (Wi-Fi) to gain unauthorized access, intercept data, or launch attacks against connected devices.

Ransomware

Malware that encrypts files on a victim's computer, demanding a ransom payment in exchange for decrypting the files. Ransomware attacks can cause significant data loss and operational disruption.

Insider Threats

Malicious or negligent actions by authorized users within an organization, such as employees or contractors, who misuse their privileges to steal data, sabotage systems, or cause harm.

In Closing

Each type of hacking technique exploits specific vulnerabilities or weaknesses in systems, networks, or human behavior. Understanding these methods helps individuals and organizations implement appropriate security measures and defenses to protect against cyber threats.

20 ... BREACH LIBRARY: LINKS TO VIEW

You Tube Channels

- HAK5
- LiveOverflow
- Hackersploit
- John Hammond
- The Cyber Mentor

Physical Readings

- Countdown to Zero Day
- Ghost in the Wires
- Hacking: The Art of Exploitation
- The Art of Invisibility
- Social Engineering: The Science of Human Hacking

Industry Certifications

- CompTIA Security+
- Cisco CCST Cybersecurity
- Fortinet Network Security Associate

Cybersecurity Tools

- Wireshark
- NMap
- Kali Linux

- Metasploit
- Nessus
- Snort
- Burp Suite
- Bitdefender
- OpenVAS
- myEMATES

Blogging

- Krebs on Security
- Schneier on Security
- Graham Cluley
- The Hacker News
- Dark Reading

Competitions

- National Collegiate Cyber Defense Competition
- Collegiate Penetration Testing Competition
- US Cyber Challenge
- ICL Collegiate Cup
- DEF CON Capture the Flag

Forums

- Wilders Security Forums
- MalwareTips Forums
- Antionline Forums
- Bleeping Computer
- Stack Exchange - Information Security

Open-Source Projects

- Github
- OSSEC
- HackerOne
- Bugcrowd
- OWASP

Online Learning

- Crybrary
- SANS Institute
- LetsDefend
- Hack the Box
- Try Hack Me

Professional Associations

- Information Systems Security Association (ISSA)
- (ISC)²
- ISACA
- CAE Community
- Cybersecurity and Infrastructure Security Agency (CISA)

21 ... COMMON NETWORK PORTS

Legend for Ports

	Chat
	Encrypted
	Gaming
	Malicious
	Peer to Peer
	Streaming

Ports

7	Echo
19	Chargen
20-21	FTP
22	SSH/SCP
23	Telnet

25	SMTP
42	WINS Replication
43	WHOIS
49	TACACS
53	DNS
67-68	DHCP/BOOTP
69	TFTP
70	Gopher
79	Finger
80	HTTP
88	Kerberos
102	MS Exchange
110	POP3
113	Ident
119	NNTP (Usenet)
123	NTP
135	Microsoft RPC
137-139	NetBIOS

143	IMAP4
161-162	SNMP
177	XDMCP
179	BGP
201	AppleTalk
264	BGMP
318	TSP
381-383	HP Openview
389	LDAP
411-412	Direct Connect
443	HTTP over SSL
445	Microsoft DS
464	Kerberos
465	SMTP over SSL
497	Retrospect
500	ISAKMP
512	rexec
513	rlogin

514	syslog
515	LPD/LPR
520	RIP
521	RIPng (IPv6)
540	UUCP
554	RTSP
546-547	DHCPv6
560	monitor
563	NNTP over SSL
587	SMTP
591	FileMaker
593	Microsoft DCOM
631	Internet Printing
636	LDAP over SSL
639	MSDP (PIM)
646	LDP (MPLS)
691	MS Exchange
860	iSCSI

873	rsync
902	VMware Server
989-990	FTP over SSL
993	IMAP4 over SSL
995	POPs over SSL
1025	Microsoft RPC
1026-1029	Windows Messenger
1080	SOCKS Proxy
1080	MyDoom
1194	OpenVPN
1214	Kazaa
1241	Nessus
1311	Dell OpenManage
1337	WASTE
1433-1434	Microsoft SQL
1512	WINS
1589	Cisco VQP
1701	L2TP

1723	MS PPTP
1725	Steam
1741	CiscoWorks 2000
1755	MS Media Server
1812-1813	RADIUS
1863	MSN
1985	Cisco HSRP
2000	Cisco SCCP
2002	Cisco ACS
2049	NFS
2082-2083	cPanel
2100	Oracle XDB
2222	DirectAdmin
2302	Halo
2483-2484	Oracle DB
2745	Bagle.H
2967	Symantec AV
3050	Interbase DB

3074	XBOX Live
3124	HTTP Proxy
3127	MyDoom
3128	HTTP Proxy
3222	GLBP
3260	iSCSI Target
3306	MySQL
3389	Terminal Server
3689	iTunes
3690	Subversion
3724	World of Warcraft
3784-3785	Vertrilo
4333	mSQL
4444	Blaster
4664	Google Desktop
4672	eMule
4899	Radmin
5000	UPnP

5001	Slingbox
5001	iperf
5004-5005	RTP
5050	Yahoo! Messenger
5060	SIP
5190	AIMACQ
5222-5223	XMPP/jabber
5432	PostgreSQL
5500	VNC Server
5554	Sasser
5631-5632	pcAnywhere
5800	VNC over HTTP
5900+	VNC Server
6000-6001	X11
6112	Battle.net
6129	DameWare
6257	WinMX
6346-6347	Gnutella

6500	GameSpy Arcade
6566	SANE
6588	AnalogX
6665-6669	IRC
6679-6697	IRC over SSL
6699	Napster
6881-6999	BitTorrent
6891-6901	Windows Live
6970	Quicktime
7212	GhostSurf
7648-7649	CU-SeeMe
8000	Internet Radio
8080	HTTP Proxy
8086-8087	Kaspersky AV
8118	Privoxy
8200	VMware Server
8500	Adobe ColdFision
8767	TeamSpeak

8866	Bagle.B
9100	HP JetDirect
9101-9103	Bacula
9119	MXit
9800	WebDAV
9898	Dabber
9988	Rbot/Spybot
9999	Urchin
10000	Webmin
10000	BackupExec
10113-10116	NetIQ
11371	OpenPGP
12035-12036	Second Life
12345	NetBus
13720-13721	NetBackup
14567	Battlefield
15118	Dipnet/Od dbob
19226	AdminSecure

19638	Ensim
20222	Usermin
24800	Synergy
25999	Xfire
27015	Half-Life
27374	Sub7
28960	Call of Duty
31337	Back Orifice
33434+	traceroute

ABOUT THE AUTHOR

I was born in an Appalachian county in central Kentucky. I was once an At-risk, First-time, Underprepared freshman college student, and I would have surely benefited from some gathered information as this.

I received an AS, BS, and MS and was a student worker for my AS and a teaching assistant for my BS and MS degrees. I currently hold 2 undergraduate degrees and 4 graduate degrees and have taken many courses throughout the years. I do not remember having any courses on the topic of retention. Since three of my graduate-level degrees are in Education, I have had many graduate courses directly related to students: Training Materials, Methods, and Evaluations; Creative Problem Solving; Instructional Design; Psycho-educational Assessments; Analysis and Design of Educational and Instructional Systems; Management and Evaluation of Instructional Technology and Distance Education Programs; Instructional Design and Media; Theories of Learning; and Curriculum Articulation, Teaching, Technology, Renewal, and Program Development.

All of those previous graduate courses are directly related to students and their learning. Like I mentioned, my doctorate is in Education not Research. There is a difference, I believe, in both avenues of doctoral programs, but that is another discussion and has been and will be a confusing topic, somewhat possibly a tree branch related to the success of post-secondary college students. I believe successfully taking courses and training directly relating to students and their outcomes has a bearing on how educators relate and come across to students. I believe this to be true. I can't believe anyone in the educational realm is not giving contrary thought to this.

Years later, I am involved in post-secondary education, bringing my ideas, concepts, philosophies, and experiences to my students. Students are the sole reason I do what I do. I will repeat once again: If one single student benefits, I have done my job and fulfilled the purpose of the writing. Currently, I am in the College of Business, Engineering and Technology where I am the Acting Chair and Assistant Professor of the School of Mathematics and Computer Science at Kentucky State University, where my specialties are Cybersecurity, Information Technology, and Network Engineering.

NOTES

The End

www.ingramcontent.com/pod-product-compliance
Lightning Source LLC
Chambersburg PA
CBHW051419090426
42737CB00014B/2749